Precise-Rewritten

Exact Values

in

Trigonometry

Bhava Nath Dahal

This book on series with:

Exact Values in Trigonometry: Five New Techniques

    Publisher: CreateSpace Independent Publishing Platform

    ISBN-10: 1536995002

    ISBN-13: 978-1536995008

    Kindle ASIN B01KFP3560

Series - 2

Copyright © 2016 Bhava Nath Dahal

All rights reserved.

ISBN-13: 978-1537299297

ISBN-10: 1537299298

My words

In the response of publication of **Exact Values in Trigonometry: Five New Techniques**, few of the reader requested to explain the individual methods described in that book, especially Arc-Line method and Precise-Rewritten method. There two methods are easy for calculation with exact values and hence became popular comparing than other method. Consequently, I decided to explain each of five methods separately. This is why, this book became part of series of **Exact Values in Trigonometry: Five New Techniques.**

One point require disclosing without any hesitation is that I am not a mathematician. Therefore, this is not a professional assignment, rather just an amateur interest. Nevertheless, this is new concept and possibly so simple too. Every step of study, writing, typing, formatting and whatever required done by single hand. So, any of the shortfall in this series of mine, with my apologies. For direct access to me, I request professional commentaries through bhavanathdahal@gmail.com.

Bhava Nath Dahal

Kathmandu, Nepal

2016 August 25 (Janmastami)

To

Dāsu Bhātta, Shivāram Pādhyā, Bashudev Pādhyā,
Benumadhāv Pādhyā, Tulāram Pādhyā, Shankār Pādhyā,
Kāshināth, Hirāmani, Govindā, Gobardhān,
Umāpati, Srināth, Govindānath,
Bhimnāth – Kāusalyā Dahal

Table of Content

CHAPTER 1 TECHNICAL TERMINOLOGY

1.1	Trigonometric Ratios	1
1.2	Trigonometric Formulae	3
1.3	Chord and Supplementary Chords	5
1.4	Number Theory	10
1.5	Constructible angles	14

CHAPTER 2 PRECISE-REWRITTEN METHOD

2.1	Precise-Rewritten Concept	16
2.2	Sine of any angle	33
2.3	Exact Values of integer angles	42
2.4	Exact Values of polygon chords	51
2.5	Proof for above values are exact	54
2.6	Constants for some nested radicals	58

Chapter 1
Technical Terminology

1.1 Trigonometric Ratios

The length of perpendicular from the vertex of an angle of a triangle to the opposite base is Sine of opposite angle of that perpendicular. In the modern mathematics, this is deemed as ratio of perpendicular to hypotenuse of a right angle triangle. Based on modern perception, there are six trigonometric ratios- Sine, Cosine, Tangent, Cotangent, Secant and Cosecant for an angle. For written text, they are written as Sin A, Cos A, Tan A, Cot A, Sec A, CSC A for an angle named as A. Many people use variations mainly on CSC A as Cosec A or use 2-digit notations for all trigonometric relations.

Apart from above six main ratios, there are less-common-use 10-ratios as Versine (Utkrāmajyā), Vercosine, Coversine, Covercosine, Exsecant, Excosecant, Haversine, Havercosine, Hacoversine and Hacovercosine.

There are many types of method to find the value of above trigonometric ratios. For example, geometric method gives trigonometric values of few angles. Algebraic formulae cover more angles. For trigonometric ratios of each and every angles, interpolation is major method. Using varies interpolation techniques, trigonometric table, computer programs and calculators having trigonometric function available now.

In this book, exact value of an angle has prescribed using Chord (a) of an angle, its supplementary chord (b) or their product (ab) make all of the trigonometric ratios under Precise-Rewritten method. The meaning and relationship of chord (a), supplementary chord (b) has described in sub-chapter 1.3.

1.2 Trigonometric Formulae

Based on academic, technical and practical use, there are six main trigonometric ratios. There are variety of ways to establish the relations. As we already discussed, Precise-Rewritten method of exact value of trigonometry is based on chord (a), supplementary chord (b) and their product (ab).

Main relations:

    i. $\sin A = ab/2$
    ii. $\csc A = 2/ab$
    iii. $\cos A = 1 - a^2/2$
    iv. $\sec A = 2/(2-a^2)$
    v. $\tan A = ab/(2-a^2)$
    vi. $\cot A = (2-a^2)/ab$

Verse and Coverse relations:

Practically in less use, there are further six trigonometric ratios. They are two in each sector of Verse-based, Coverse-based and Ex-based. Their chord-based formulae are as follows:

    vii. $\text{Versin } A = a^2/2$
    viii. $\text{Vercos } A = b^2/2$
    ix. $\text{Coversin } A = 1 - ab/2$
    x. $\text{Covercos } A = 1 - ab/2$
    xi. $\text{Exsec } A = a^2/(2-a^2)$
    xii. $\text{Excsc } A = (2-ab)/ab$

Haverse and Hacoverse relations:

Practically almost no use, there are further four trigonometric ratios. They are half of Verse-based and Coverse-based formula. Their chord-based formulae are as follows:

    xiii.   Haversine $= a^2/4$

    xiv.   Havercosine $= b^2/4$

    xv.   Hacoversine $= (2-ab)/4$

    xvi.   Hacovercosine $= (2+ab)/4$

If we know the length of chord, half of that chord ($a/2$) will be the Sine of half of angle representing that chord (not included above). This is the magical development of determination of Sine of an angle since last 2000 years.

1.3 Chord and Supplementary Chords

In the conceptual study for Precise-Rewritten method, some of the trigonometric terminologies need to be understand. For generic study, very few of this concept requires.

Chord and Supplementary Chord

The line that passes through two points of circumference of a circle is known as chord. Chord of an angle at the center and its supplementary chord are perpendicular each other. Consequently, sum of squares of chord and its supplementary chord is four. In this book, chord of an angle is denoted as 'a' and its supplementary chord is denoted as 'b'. In the circle below,

∠AOB=A˙,

Length AB is chord for ∠A=a,

Length BD is supplementary chord of for ∠A=b,

There are certain relations regarding both chords, a and b. Based on Guru Pythagoras's theorem for perpendicular chord (a) and supplementary chord (b), $b = \sqrt{4 - a^2}$ and vice-versa.

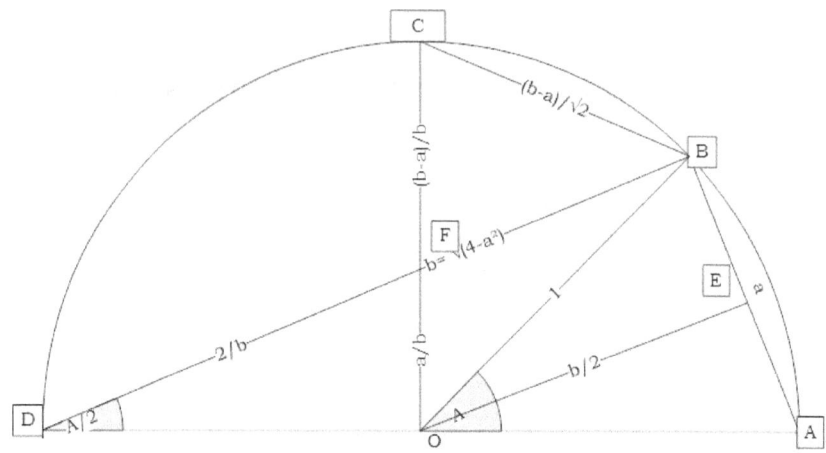

Scholars in trigonometry know the meaning of chord. They could have used chord (a) within a circle. Under Precise-Rewritten method, use of supplementary chord (b) is an additional contribution. Another interesting point for supplementary chord (b) under Precise-Rewritten method, it is just a change in sign after the first radical. For example, chord (a) for 22.5° is $\sqrt{(2 - \sqrt{2})}$, its supplementary chord (b) will be just $\sqrt{(2 + \sqrt{2})}$. For 5.625°, chord (a) and supplementary chord (b) will be $\sqrt{(2 - \sqrt{(2 + \sqrt{(2+ \sqrt{2})})})}$ and $\sqrt{(2 + \sqrt{(2 + \sqrt{(2+ \sqrt{2})})})}$. We can easily multiply these 'a' and 'b' for another contributory 'ab'. Using $(a + b)(a - b) = a^2 - b^2$, the product 'ab' for 22.5° is $\sqrt{(four - 2)} = \sqrt{2}$ and for 5.625° is $\sqrt{(2 - \sqrt{(2+ \sqrt{2})}}]$.

For an easy and catchy formula, in the last of each formula, we have used ']' in the place of numerous closing brackets throughout this book. In the real calculation, we need to use actual brackets as simple mathematics.

Half angle Chords

Chord of a half-angle is one of the contributory for Precise-Rewritten method. We use half-angles (called here as central) of route – 90°.

Square root of the difference of 2 and supplementary chord of that angle is chord of half angle.

Square root of the sum of 2 and supplementary chord of that angle is chord of complementary of half angle.

Crd. A/2 $= \sqrt{2-b}$

Crd. (90 − A/2) $= \sqrt{2+b}$

Chord of double angle is the product of chord and supplementary chord of that angle.

Crd. 2A $= ab$

Double angle chord is just 'b' times 'a' of an angle.

Central and its Chords

Precise-Rewritten method is critically based on *centrals*. Half of 90° i.e. 45° is the first central. Half of the first central is second central; therefore, 22.5° is the second central. In each step of calculation, half of earlier central is current central. We convert the given angle into centrals (rewritten). Basic central for Precise-Rewritten is 90°, but it is not a rewritten central for the calculation process. Let us see example of central and rewritten for angles.

33.75° = 45° − 22.5° + 11.25° = 45° − (22.5° − (11.25°]

56.25° = 45° + 22.5° − 11.25° = 45° + (22.5° − (11.25°]

78.75° = 45° + 22.5° + 11.25° = 45° + (22.5° + (11.25°]

11.25° = 45° − 22.5° − 11.25° = 45° − (22.5° + (11.25°]

All angles in above examples are rewritten in centrals with bracket in each step.

Centrals may compute from either measurement of angle. Degree (°) has taken as basis for this book. Users may easily convert or prepare the centrals in other measurement system. For example, in the French Degree (^{g}) system, centrals will be in form of 50^g, 25^g, 12.5 ^{g}, ... In the circular measurement (radian ^{c}), centrals will be in form of π/4, π/8, π/16, ... In Hindu measurement (*āmsa*) system, same as English Degree (°) is the form of centrals.

Binary-sign techniques

This is the sign determination of angle – rewritten for a given angle. If we observed 33.75°, it is based on centrals as 45° – 22.5° + 11.25°. In this case, 33.75° has converted into centrals in half-series of 90°. For chord conversion from these centrals, we need to rewrite these angles in bracket form as 45° – (22.5° + (11.25°)).

For an angle like 18.984375°, angle-rewritten as this model is a complex task. For making simple, binary-sign techniques of angle-rewritten designed as:

Central	Angle (A)	Sign
90		
45	45	2
-22.5	22.5	-2
-11.25	11.25	2
5.625	16.875	-2
2.8125	19.6875	2
-1.40625	18.28125	-2
0.703125	18.984375	-2

PRECISE-REWRITTEN METHOD

In this case, our target angle is 18.984375°. Starting from the first central (45°) towards to target angle, second central (22.5°) requires subtracting; third central (11.25°) requires subtracting; fourth central (5.625°) requires adding; fifth central (1.40625°) requires subtracting and sixth central (0.703135°) requires adding.

In each step of centrals, the second is half of the first; therefore, the quotient is always 2. However, there is either +2 or -2 based on sign of centrals. For example, 90/45= +2, 45/ (-22.5)= -2 etc. Alternatively, if the current central has added into the cumulative, that is '+2', or if it is reducing the cumulative, '-2'. This is how, angle-rewritten made simple calculation. The result is always ±2, and hence the technique for rewriting an angle is *binary-sign technique*.

1.4 Number Theory

Few of the terminologies in this book are relating to number theory.

Surds or irrational numbers

We knew, $\sqrt{1}$ or $\sqrt{4}$ or $\sqrt{2.25}$ have rational roots of 1, 2 or 1.5. What is the case for $\sqrt{2}$ or $\sqrt{3}$? Our root is never ending calculation nor repeating in nature. Therefore, these are surds.

A root of number that has not a rational number is surds. Surds are irrational number. In square root or nested radicals, we face many irrational numbers. Development and definition of surds goes to credit of Guru Al-Khwarizmi (780-850) as inaudible (deaf or mute) number in Persian. Latter the meaning of deaf had translated into Latin as *surdus* (deaf or mute) and modern name as *Surds*. Almost all trigonometric ratios are irrational in nature.

Precise-Rewritten method solely depends upon $\sqrt{2}$ and nested radicals along with $\sqrt{2}$. Therefore, each and every trigonometric ratios will be in irrational format.

Finite and Infinitive Series of Centrals

Angle that can be rewritten with finite number of centrals is finite-series of centrals. For example, $16.875°$ can be rewritten as 45-22.5-11.25+5.625; therefore, $16.875°$ is a member of finite series of centrals. There are four centrals and the chord (a) or supplementary chord (b) will have four $\sqrt{2}$'s with nested radicals.

There are other angles, which cannot be rewritten in simple forms of centrals, need to expand into infinite number of calculations. These are infinitive series of centrals. For $17°$, we cannot limit the calculation for 4 or 5 centrals. We need to compute infinitive centrals to arrive 17°. This is the member of infinitive series of centrals.

Result under Precise-Rewritten method for infinitive series of centrals are of two types: one has repeating radicals and another non-repeating radical. Integer angles and polygons have repeating radicals. Do not be worried, we need not compute infinitive centrals for trigonometric ratios. We shall find repeating nested radicals during calculations.

Nested radicals

Root of a number with further root over it is known as nested radical. For example, in $\sqrt{(2 - \sqrt{2})}$, second $\sqrt{2}$ is nested by first [square] root. In $\sqrt{(2 - \sqrt{(2 + \sqrt{(2 - \sqrt{2})})}}$, last $\sqrt{2}$ is nested by three more [square] roots. Precise-Rewritten method solely written in the format of nested radicals. Each central has, just and just, converted into nested radicals of 2 as $\sqrt{(2 - \sqrt{2})}$ for 45°-22.5°. This $\sqrt{(2 - \sqrt{2})}$ is chord for 45°.

Number of nested radicals in any trigonometric ratio depend upon the appropriate accuracy or precision for the accuracy. If the nested radicals is not repeating in nature, more nested radicals for more accuracy requires.

Sum of Nested radicals

For any angle except angles within definitive series, the chord calculated under Precise-Rewritten method has special characteristics of repetition of right-most nested radicals' upto infinitive. In case of definitive series, there might be repeating nested radical, but it limits up to defined numbers only.

For example, under Precise-Rewritten method, chord for 60° is $\sqrt{(2-\sqrt{(2-\sqrt{(2-\sqrt{(2\ldots}}}}$ in place of classical chord length of 1. In this chord, right-most $\sqrt{(2}$ has repeated infinitely. This is infinite series as $\sqrt{(2+\overline{-\sqrt{(2]}}}$.

Repeating nested radicals have special characteristics.

In the following list, few numbers are calculated to find sum of nested radicals. Carefully observing, the sum of infinitive radicals of a number (n) has found as product of that number and just lower number [n = S×(S-1)]. For example, sum of nested square root is calculated as 3 in the right hand side for n=6. We knew the sum S=3, the input n must be 3×2= 6. For sum of 5, the input must be 5×4= 20. For sum of 7, the input must be 7×6= 42.

n	$S = \sqrt{(n+\sqrt{(n+\sqrt{(n+\sqrt{(n+\ldots\ldots}}}}$
2	2
6	3
12	4
20	5
30	6
42	7

Formulating above example for unknown sum (S) for known input (n):

S (S-1)= n

Or, $S^2 - S - n = 0$

Solving for S,

$S = \frac{1}{2}[1+\sqrt{(1+4n)}]$

This is Guru Ramanujan's (22 December 1887 – 26 April 1920) formula for nested radicals with little adjustment. For subtraction, the formula for will be:

$$S = \tfrac{1}{2}\left[-1+\sqrt{(1+4n)}\right]$$

Taking this formula, we need to resolve the repetitive part of each nested radicals in the chord for decimal representation of trigonometric values derived under Precise-Rewritten method.

Applying above formula of sum of nested radicals for n=2, we shall receive chord for 60° is 1 as we known since long ago.

1.5 Constructible angles

From Precise-Rewritten method, we can compute trigonometric ratios of all angles (integer or not). Irrespective of their classical chord length, we use nested radical of √2 in each case. We know √2 is constructible. Not only √2, but also any combination of √2 like √(2 − √2) or (√2+√2 - √2) etc. can be constructible using compass and straightedge only. A compass and straightedge can easily bisect a straight-line as well as an angle with little knowledge of geometry. Our initial point of start is 45° as the first central; and each step new central is just half of earlier. Hence, each and every angle under Precise-Rewritten method is constructible.

Does this mean we can construct any regular polygon using compass and straightedge only? However, the answer is 'theoretically no'. Gauss–Wantzel theorem by Guru Carl Friedrich Gauss (1777-1855) and Guru Pierre Wantzel (1814-1848) is still valid for constructible polygons. Precise-Rewritten method prescribes for more and more inventory of constructible lines within a circle. It recommends exact values based on repeating nested radicals. Only the nested radicals (of √2) which has *definite series* can be 'theoretically' constructible. An angle having chord with nested radicals with infinitive repeat or no repeat up to infinitive steps cannot be constructible 'theoretically'.

For example, it prescribes exact value of all integer angles like Sin 1° or Cos 1°; Sin 2° or Cos 2°. However, these are in infinitive series. For construction a line, there should be end-point. Therefore, even, we have exact values for Crd. 1° or Crd. 2°, we cannot construct a polygon based on this.

For the chord determined under Precise-Rewritten method, we can construct chord for 40° having n-digits of accuracy but not 40° itself. We can construct a line for 40.000….. *[billion or trillions 0's]*……. degree, but still we will have reminder after those billions or trillions zero's. Therefore, Precise-Rewritten method gives *precisely exact values* as the name of method suggests.

Practically, the scenario is different. We can divide the solar-orbit in exact 9 pieces (9-gon) for precision level of 16-digits after decimal for 40°. For practical purpose, we need not to calculate the precision accuracy for large numbers like billion digits of trillion-digits.

Chapter 2
PRECISE-REWRITTEN METHOD

2.1 Precise-Rewritten Concept

Classically, we know few angles having exact trigonometric values. After developing new method, we added more angles having exact trigonometric values. However, we never discuss back for definition of 'exact trigonometric value'. Even we say that exact value of trigonometric ratio of an angle; in fact, there is no exact values. All the values of Sine of an angle except Sine 0°, 30° and 90° are not exact value. For the remaining cases of Sine of an angle, exact value means exact in the notation only, not in the number of measurement itself.

For example, well-known value for Sin 60° is √3/2 under radical notation. It seems as exact value, or it is said to be exact value. However, in real number system (decimal system), we should relay on the precision of digits after decimals, because √3 itself is an irrational number[1]. Similar

---

[1] Guru Theodorus (5th Century BC) proved √3 is an irrational. For his respect, it is called as Theodorus' constant.

the case for Sin 45° or 36°. All these angles, which have so-called exact value of trigonometric ratios, have a value just near to the exact in decimal system. As a result, all the remaining cases of ultimate value of Sine of an angle, except 0°, 30°, 90°, is always irrational. There are similar irrationality for other trigonometric ratios. In this book, for calculative purpose and for explanation, Sine of an angle has taken as base.

We shall see the chord or supplementary chord of each angle will be depend upon either $\sqrt{2}$ or nested radicals. However, $\sqrt{2}$ itself is an irrational number and its nested radical obviously result irrational values in the result. That lead all the trigonometric values are irrational or surds. Coming back to value of π, it is still counting for millions digits and will be counted further in the future. Similar to the case for $\sqrt{2}$ or $\sqrt{3}$, the calculated accuracy of digits after decimals are increasing in geometric ratio. In the future further accuracy will be set up.

In other words, in the decimal system of number, trigonometric ratios of an angle is that number which do have *appropriate accuracy* for the user. Under this concept of *appropriate accuracy*, we can calculate exact value of any angle using *Precise-Rewritten method*. This method determines exact values of most angles then we have now, however still the precision requires for real measurement. Hence, the method contains the word *'Precise'*. For many angles, it recommends 'exact value' rather than 'precise value'.

For determination of trigonometric ratios, an angle (A) is rewritten under

Binary-sign techniques (see sub-chapter 1.3). This is why the method is named as Precise-Rewritten method.

Accuracy

For *any angle including classically known trigonometric values*, accuracy under Precise-Rewritten method has two fold- one is exact and another is based on the user-defined precision level. For a polygon or for an integer angle, there is 'exact value'. For other angles, user-defied precision is requires.

As already explained, either √2, √3 or √5 is an irrational number. For practical purpose, we should set some digits after decimal for √2, √3 or √5. Recommended appropriate accuracy under Precise-Rewritten method may be same number of digits of accuracy we accept for √2, √3 or √5. As a result, so-called exact value or user-defined precision produces same result in decimal presentation.

For example, we may use 1.732050807568880 for √3 and 0.866025403784439 for √3/2 having 15-digits after decimal point. This is 1.732 05080 7568 8772 9352 7446 3415 05872 36694 2805 2538 1038 0628 0558 0… in the Sequence A002194 in the OEIS. Carefully observing on these three figures, the first is not exactly double of the second nor half of the third. This is the *appropriate accuracy* accepted by the user in case of √3.

Under Precise-Rewritten method, Sin 60° is √(2 + √(2 -√(2 -√(2 -√(2 -√(2 -√(2 -√(2 -√(2 -√(2 -√(2 -√(2 -√(2 -√(2 -√(2 -√(2 -√(2 -√(2 -√(2 -√(2 -√(2 -√(2 -√(2 -√(2 -√(2 -√(2 -√(2 -√(2 -√(2 -√(2 -√(2 -√(2 -√(2 -√(2 -√(2 -√(2 -√(2 -√(2 -√(2 -√(2 -√(2 -√2]/2. Its decimal value[2] is 0.866025403784439, which is same as above[3].

Another example of 20/3 is 0.6666666$_6$.... Multiplying back with 3, the exact value is something other than 20. Similar the case with 1/3 and back multiplication by 3. In these cases, appropriate accuracy for 1 is 0. 99999$_9$ ... and vice versa. Nevertheless, for practical purpose, we cannot speak or write infinitive digits; rather we set a fix-length; e.g. 5-digits, 12-digits, 15-digits, 25-digits etc. In the computer programming also, programmers set significant digits (SD). In that case, we assumed n numbers of 9's after decimal is 1. This is the user-defined *appropriate accuracy (Precise accuracy)* for our purpose. However, Precise-Rewritten method prescribed a big constellation of 'exact values', *appropriate accuracy* is most required concept for practical purpose.

Precise accuracy have two input parameters. The first is the number of digits handled during each process of calculation and the second is the number of calculations itself. If we define the appropriate accuracy of 15-digits, we must use, *at least,* 15-digits after decimal in each step of calculation.

For centrals (means half-angles from 90°, see above), when we arrive at the 15-digits after decimal, our *maximum error* will be half of 15th digit[4]. For a sum of nested radical of 2,

---

[2] For an infinitive series with nested radicals for 2 in form of √(2 - √(2 - .... is equal to 1 and √(2 +√(2 +.... is equal to 2. In the above case Sin 60° has value of √(2 + √(2 -√(2 - ...]/2, which is, in fact, √(2 +1)/2 = √3/2. Therefore, for an infinitive series under Precise-Rewritten method, trigonometric ratios will be exact value.

[3] To make the presentation easy, all the closing brackets are collapsed by single bracket ']' in this book.

[4] Using, sum of series of powers for $1/n = \frac{1}{n-1}$ for n=2; our precision will have maximum difference of 0.5. This maximum will occur, if we

our *maximum error* will be 15th root[5] $\sqrt{2}$. In general, *maximum error* is same as the ending point of precision of original input.

For example of *maximum error*, if we set 15-digits accuracy, we should take *at least* 15-digits during calculation. The difference after those 15-digits may be maximum of half of 15th digit; e.g. 23.4±0.0000000000000005 for 23.4. However, there were numerous '+' or '–' within our calculation until 15th step, the difference is less than ±0.0000000000000005. Similar the case of chord of 23.4°. Precise-Rewritten method have only nested $\sqrt{2}$. Consequently, the maximum error in Sin 23.4° will be a half of 15th root of 2. Again, there are numerous + or – within our case, this error is less than a half of maximum error of 15th root of 2.

To reduce this gap, of course, we have two options:

Firstly, should increase the calculation process vis-à-vis no. of digits in each step. Roughly, 10 further calculations increases the appropriate accuracy by 3-digits. Therefore, to obtain accuracy of 100-digits, we should handle *at least* 100-digits after decimal in each step and have around 300 nested radicals. At that point *maximum error* will account of a half of $\sqrt[300]{2}$.

Secondly, we may take sum of repeating nested radicals. In this case, the result will an irrational number but not a countable error. However, this alternative has limitation. The repeating nested radicals is possible for an angle (A) which has 360/A is a natural integer.

Lastly, if someone need not decimal values, exact values in the nested radical will be in hand.

---

subtract or add the halves infinitively. If we have used + or – sign during calculations, the differences fall less than half of last digit.

[5] Using, sum of series of nested radicals for 2, the sum is 2 for $\sqrt{(2+\sqrt{(2+....}}$. because x (x-1) = 2, if and only if x=2. Therefore, in the nth calculation step, this error will accumulated as n^{th} root of 2.

Inputs

If above concept and terminologies are somewhat familiar, for Precise-Rewritten method, we need three generic inputs:

> Firstly, degree measure of angle (we can change it into another measure too);
>
> Secondly, *appropriate accuracy* of precision (say 10-digits accurate, 20-digits accurate, or n-digits accurate); and
>
> Thirdly, Centrals (half-series under route-90°) under binary-sign technique.

With regards inputs, we need not discuss at all. First two are extremely generic in nature. The third seems a bit technical, may learn within few minutes or simple learning is enough in examples below.

Based on above three inputs, we can compute chord (a) of any angle. From this chord (a), we can compute the supplementary chord (b) or their product (ab). From 'a', 'b' or 'ab' we compute all trigonometric ratios of *any angle* using the simple arithmetical process. After few exercise, student at elementary school who has general knowledge of trigonometry can find the exact trigonometric value for any angle.

Mathematical process

> Binary-sign technique is the basis for determination of chord of given angle (A). In fact, above process of binary-sign technique gives the chord of double angle (Crd. 2A). Half of this chord is Sine of given angle (Sin A).

Let us take above example of 18.984375° for explain the mathematical process of Precise-Rewritten method.

Calling back the example of binary-sign, replace each 2 (binary) by nested radical [√(2)] with sign of binary-sign. The result is *always* chord of double angle.

Angle (A)	Sign	Crd 2A
45	2	$\sqrt{2}]$
22.5	-2	$\sqrt{2-\sqrt{2}}]$
11.25	2	$\sqrt{2-\sqrt{2+\sqrt{2}}}]$
16.875	-2	$\sqrt{2-\sqrt{2+\sqrt{2-\sqrt{2}}}}]$
19.6875	2	$\sqrt{2-\sqrt{2+\sqrt{2-\sqrt{2+\sqrt{2}}}}}]$
18.28125]	-2	$\sqrt{2-\sqrt{2+\sqrt{2-\sqrt{2+\sqrt{2-\sqrt{2}}}}}}$
18.984375 $-\sqrt{2}]$	-2	$\sqrt{2-\sqrt{2+\sqrt{2-\sqrt{2+\sqrt{2-\sqrt{2}}}}}}$

In each step, chord of double angle (2A) has given above though our target angle was 18.984375°. Each binary has converted into nested radical with sign is the chord (a) of double angle. So,

Angle (A)	2A	Sign	Chord 2A
45	90	2	$\sqrt{2}]$
22.5	45	-2	$\sqrt{2-\sqrt{2}}]$
11.25	22.5	2	$\sqrt{2-\sqrt{2+\sqrt{2}}}]$
16.875	33.75	-2	$\sqrt{2-\sqrt{2+\sqrt{2-\sqrt{2}}}}]$
19.6875]	39.375	2	$\sqrt{2-\sqrt{2+\sqrt{2-\sqrt{2+\sqrt{2}}}}}$
18.28125 $-\sqrt{2}]$	36.5625	-2	$\sqrt{2-\sqrt{2+\sqrt{2-\sqrt{2+\sqrt{2}}}}}$
18.984375	37.96875	-2	$\sqrt{2-\sqrt{2+\sqrt{2-\sqrt{2+\sqrt{2}}}}}$ $-\sqrt{2-\sqrt{2}}]$

Respectful Greek and Hindu Gurus gifted us that the half of chord is Sine of half angle, since before 2000 years.

PRECISE-REWRITTEN METHOD

Therefore, half of chord of each step ($\frac{1}{2} \times$ chord 2A) is Sin A in each step.

From above chord of double angle, we can compute supplementary chord of double angle. For this sign after of first radical has to be changed into '+'. Our target angle was lower than 45°, therefore, we use '-' sign after first radical. Supplementary chord will be higher than 45°, so we need to add and need '+' sign. Hence, chord (a) and supplementary chord (b) of each angles will be as follows:

Angle	'a'	'b'
90	$\sqrt{2}$]	$\sqrt{2}$]
45	$\sqrt{(2 - \sqrt{2}}$]	$\sqrt{(2 + \sqrt{2}}$]
22.5	$\sqrt{(2 - \sqrt{(2 + \sqrt{2}}}$]	$\sqrt{(2 + \sqrt{(2 + \sqrt{2}}}$]
33.75	$\sqrt{(2 - \sqrt{(2 + \sqrt{(2 - \sqrt{2}}}}$]	$\sqrt{(2 + \sqrt{(2 + \sqrt{(2 - \sqrt{2}}}}$]
39.375	$\sqrt{(2 - \sqrt{(2 + \sqrt{(2 - \sqrt{(2 + \sqrt{2}}}}}$]	$\sqrt{(2 + \sqrt{(2 + \sqrt{(2 - \sqrt{(2 + \sqrt{2}}}}}$]
36.5625	$\sqrt{(2 - \sqrt{(2 + \sqrt{(2}}}$ $-\sqrt{(2 + \sqrt{(2 - \sqrt{2}}}$]	$\sqrt{(2 + \sqrt{(2 + \sqrt{(2}}}$ $-$ $\sqrt{(2 + \sqrt{(2 - \sqrt{2}}}$]
37.96875	$\sqrt{(2 - \sqrt{(2 + \sqrt{(2}}}$ $-\sqrt{(2 + \sqrt{(2 - \sqrt{(2 - \sqrt{2}}}}$]	$\sqrt{(2 + \sqrt{(2 + \sqrt{(2}}}$ $-$ $\sqrt{(2 + \sqrt{(2 - \sqrt{(2 - \sqrt{2}}}}$]

We have listed sixteen trigonometric ratios including six main ratios in sub-chapter 1.2. All of them were depended upon of either 'a', b or ab. Now, we have all three inputs and can compute all sixteen trigonometric ratios.

In this method, one need to perform just simple arithmetical process of Angle-Rewritten for route-90° until achieving the target angle (or target angle with *appropriate accuracy*). In case there is no reminder after achieving target angle, appropriate accuracy is 100% or exact value. In the above case, our target angle(s) is 18.984375° or its doubles. In the binary-sign technique, there is no reminder. Therefore, the resulting chord, supplementary chord or trigonometric ratios are 'exact value'.

Example of Sin 1° (Crd. 2°)

Let us take an example of Sin 1° for further explaining Precise-Rewritten method as a complete process.

- Make a table with 3 columns:
    o First column contains centrals (*90 and its half*) in each step. The sign has assigned based on target angle (1° in our case). If the earlier cumulative value (in second column) is higher than target 1°, the sign will be –ve; if the earlier cumulative value is lower than the target, the sign will be +ve.
    o Second column is cumulative angle trending towards target angle (1°). We should care the cumulative value, whether achieved at *appropriate accuracy* or not. Any digits after precise no. of digits accuracy may remove or round.
    o Third column is the ratio of first column and must be ±2 in each steps. Its *sign is critical* for chord (a). In our case, 90/45= +2 is first sign of first central; 45/(-22.5)= - 2 is second sign and so on.

Centrals	Target angle (A)	Sign
90.00000000000000		
45.00000000000000	45.00000000000000	2
-22.50000000000000	22.50000000000000	-2

PRECISE-REWRITTEN METHOD

-11.25000000000000	11.25000000000000	2
-5.62500000000000	5.62500000000000	2
-2.81250000000000	2.81250000000000	2
-1.40625000000000	1.40625000000000	2
-0.70312500000000	0.70312500000000	2
0.35156250000000	1.05468750000000	-2
-0.17578125000000	0.87890625000000	-2
0.08789062500000	0.96679687500000	-2
0.04394531250000	1.01074218750000	2
-0.02197265625000	0.98876953125000	-2
0.01098632812500	0.99975585937500	-2
0.00549316406250	1.00524902343750	2
-0.00274658203125	1.00250244140625	-2
-0.00137329101563	1.00112915039062	2
-0.00068664550781	1.00044250488281	2
-0.00034332275391	1.00009918212890	2
-0.00017166137695	0.99992752075195	2
0.00008583068848	1.00001335144042	-2
-0.00004291534424	0.99997043609619	-2
0.00002145767212	0.99999189376831	-2
0.00001072883606	1.00000262260437	2
-0.00000536441803	0.99999725818634	-2
0.00000268220901	0.99999994039536	-2
0.00000134110451	1.00000128149986	2
-0.00000067055225	1.00000061094760	-2
-0.00000033527613	1.00000027567148	2
-0.00000016763806	1.00000010803341	2
-0.00000008381903	1.00000002421438	2

-0.00000004190952	0.99999998230487	2
0.00000002095476	1.00000000325962	-2
-0.00000001047738	0.99999999278225	-2
0.00000000523869	0.99999999802094	-2
0.00000000261934	1.00000000064028	2
-0.00000000130967	0.99999999933061	-2
0.00000000065484	0.99999999998545	-2
0.00000000032742	1.00000000031287	2
-0.00000000016371	1.00000000014916	-2
-0.00000000008185	1.00000000006730	2
-0.00000000004093	1.00000000002638	2
-0.00000000002046	1.00000000000591	2
-0.00000000001023	0.99999999999568	2
0.00000000000512	1.00000000000080	-2
-0.00000000000256	0.99999999999824	-2
0.00000000000128	0.99999999999952	-2
0.00000000000064	1.00000000000016	2
-0.00000000000032	0.99999999999984	-2
0.00000000000016	1.000000000000000	-2

For writing Sine of target angle, start from first 2 (always +ve). Each of ±2's should be written as nested radical as $\pm\sqrt{(2}$. The sign before each nested '$\sqrt{(2}$' should be same as above given in the table. In our case

Sin 1° = $\sqrt{(2} -\sqrt{(2} + \sqrt{(2} + \sqrt{(2} + \sqrt{(2} + \sqrt{(2} + \sqrt{(2} -\sqrt{(2} -\sqrt{(2} -\sqrt{(2} + \sqrt{(2} -\sqrt{(2} -\sqrt{(2} + \sqrt{(2} -\sqrt{(2} + \sqrt{(2} + \sqrt{(2} + \sqrt{(2} + \sqrt{(2} - \sqrt{(2} -\sqrt{(2} -\sqrt{(2} + \sqrt{(2} -\sqrt{(2} -\sqrt{(2} + \sqrt{(2} -\sqrt{(2} + \sqrt{(2} + \sqrt{(2} + \sqrt{(2} + \sqrt{(2} -\sqrt{(2} -\sqrt{(2} -\sqrt{(2} + \sqrt{(2} -\sqrt{(2} -\sqrt{(2} + \sqrt{(2} -\sqrt{(2} + \sqrt{(2} + \sqrt{(2} + \sqrt{(2} + \sqrt{(2} -\sqrt{(2} -\sqrt{(2} -\sqrt{(2} + \sqrt{(2} -\sqrt{(2} -\sqrt{2} \,]/2$

PRECISE-REWRITTEN METHOD

Since the angle (1°) is accurate for 15-digits of decimal, the accuracy of Sin 1° is same as 15-digits of decimal. Before dividing by 2, the result is chord of 2°.

Exact value of Sin 1° (Crd. 2°)

In above case, accuracy is 15-digits after decimal point. There is still reminder after 15-digits of 0's in the 15th position. What is the exact value for 1°? For this, we need to careful observation of the available radicals.

The resulting radicals can be rearranged as:

Sin 1°= √(2 -√(2 +√(2

+√(2 +√(2 +√(2 +√(2 -√(2 -√(2 -√(2 +√(2 -√(2 -√(2 +√(2 -√(2

+√(2 +√(2 +√(2 +√(2 -√(2 -√(2 -√(2 +√(2 -√(2 -√(2 +√(2 -√(2

+√(2 +√(2 +√(2 +√(2 -√(2 -√(2 -√(2 +√(2 -√(2 -√(2 +√(2 -√(2

+√(2 +√(2 +√(2 +√(2 -√(2 -√(2 -√(2 +√(2 -√(2 -√(2 +√(2 -√(2

+√(2 +√(2 +√(2]/2

If we observe the table above, first line is repeating constantly. Therefore, Sin 1°= ½ √(2 -√(2 +√(2

+√(2 + √(2 +√(2 +√(2 -√(2 -√(2 -√(2 +√(2 -√(2 -√(2 ++√(2 - +√(2]

[Please care the sign of <u>repetition</u> has <u>underlined</u>.]

Without any precision for appropriate accuracy, this is *exact value* of chord for 2° and hence *exact value of Sin 1°*.

Note:

1. *The figure after 15 digits has not shown in the table, but has taken into calculation base. Therefore, calculation of each step, figure may not match exactly as we compute by calculator.*

2. *The technique to exhibit numbers here is Binary-Sign technique as described in Angle-Rewritten method in Exact Values in Trigonometry; Five New Techniques (ISBN 978-1536995008) written by same author. Even recommended, being handling sign is extremely easy under Binary-sign technique; readers may use other techniques developed in Angle-Rewritten method as well.*

3. *The result before dividing by 2 is chord (a) of 2°. Supplementary chord (b) for 2° shall be exactly same as 'a' replacing first sign '-' by '+'.*

   From availability of both chords of 2°, now we can compute value of Sine of all integer angles using classical methods. Because classical methods could give exact values of all integer angles divisible by 3. Precise-Rewritten method, until this point added 2° in the inventory. So using classical formula, we can compute Sine of all angles. Below in this chapter, Sine of all angles are given based on Precise-Rewritten method.

In the table given below, we have computed Sine of angles from 1° to 45°. The angles after 45 are complementary angles for those angles. We can compute Sine of those complementary angles using supplementary chord or just changing the first sign into '+'. For example,

Complementary angle of 1° is 89°. Therefore, in exact terms, Sin 89°= ½ √(2 -√(2 +√(2

+√(2 + √(2 +√(2 +√(2 -√(2 -√(2 -√(2 +√(2 -√(2 ++√(2 - +√(2]

[Please care the sign of <u>repetition</u> has <u>underlined</u>.]

From Sin 1°, we computed complementary angle Sin 89° too. Sin 1° and Cos 89° has same value and vice versa. From these four values, we can compute all other trigonometric ratios using classical method.

PRECISE-REWRITTEN METHOD

In sub-chapter 1.2, we have listed sixteen trigonometric formulae based on a, b, and ab. We have all these three inputs now. Hence, we can compute any trigonometric ratio for 1° and 89° with exact values.

Example of Sin 64°

Furthermore, let us take another example of Sin 64°.

		Sign
90.00000000000000		
45.00000000000000	45.00000000000000	2
22.50000000000000	67.50000000000000	2
-11.25000000000000	56.25000000000000	-2
5.62500000000000	61.87500000000000	-2
2.81250000000000	64.68750000000000	2
-1.40625000000000	63.28125000000000	-2
0.70312500000000	63.98437500000000	-2
……..	……..	….
32-steps	omitted	
……..	……..	….
0.00000000008185	63.99999999996910	-2
0.00000000004093	64.00000000001000	2
-0.00000000002046	63.99999999998950	-2
0.00000000001023	63.99999999999980	-2
0.00000000000512	64.00000000000490	2
-0.00000000000256	64.00000000000230	-2
-0.00000000000128	64.00000000000110	2
-0.00000000000064	64.00000000000040	2
-0.00000000000032	64.00000000000010	2

-0.00000000000016	63.99999999999990	2
0.00000000000008	64.000000000000000	-2

So, $\sin 64° = \sqrt{(2 + \sqrt{(2 - \sqrt{(2 - \sqrt{(2 + \sqrt{(2 - \sqrt{(2 - \sqrt{(2 + \sqrt{(2 - \sqrt{(2 + \sqrt{(2 + \sqrt{(2 + \sqrt{(2 - \sqrt{(2 - \sqrt{(2 - \sqrt{(2 + \sqrt{(2 - \sqrt{(2 - \sqrt{(2 + \sqrt{(2 + \sqrt{(2 - \sqrt{(2 - \sqrt{(2 + \sqrt{(2 - \sqrt{(2 + \sqrt{(2 + \sqrt{(2 + \sqrt{(2 + \sqrt{(2 - \sqrt{(2 - \sqrt{(2 - \sqrt{(2 + \sqrt{(2 - \sqrt{(2 - \sqrt{(2 + \sqrt{(2 - \sqrt{(2 + \sqrt{(2 + \sqrt{(2 + \sqrt{(2 + \sqrt{(2 - \sqrt{2}}/2$

Cos 64° or complementary angle, $\sin 26° = \sqrt{(2 - \sqrt{(2 - \sqrt{(2 - \sqrt{(2 + \sqrt{(2 - \sqrt{(2 - \sqrt{(2 + \sqrt{(2 - \sqrt{(2 + \sqrt{(2 + \sqrt{(2 + \sqrt{(2 + \sqrt{(2 - \sqrt{(2 - \sqrt{(2 - \sqrt{(2 + \sqrt{(2 - \sqrt{(2 - \sqrt{(2 + \sqrt{(2 - \sqrt{(2 + \sqrt{(2 + \sqrt{(2 + \sqrt{(2 - \sqrt{(2 - \sqrt{(2 - \sqrt{(2 + \sqrt{(2 - \sqrt{(2 - \sqrt{(2 + \sqrt{(2 - \sqrt{(2 + \sqrt{(2 + \sqrt{(2 + \sqrt{(2 + \sqrt{(2 + \sqrt{(2 - \sqrt{(2 - \sqrt{(2 - \sqrt{(2 + \sqrt{(2 - \sqrt{(2 - \sqrt{(2 + \sqrt{(2 - \sqrt{2}}/2$

Above example of Sin 64° is more than 45°. Its complementary angle is 26°. Only first sign will be change for Sin 26° as:

$\sin 26° = \sqrt{(2 - \sqrt{(2 - \sqrt{(2 - \sqrt{(2 + \sqrt{(2 - \sqrt{(2 - \sqrt{(2 + \sqrt{(2 - \sqrt{(2 + \sqrt{(2 + \sqrt{(2 + \sqrt{(2 - \sqrt{(2 - \sqrt{(2 - \sqrt{(2 + \sqrt{(2 - \sqrt{(2 - \sqrt{(2 + \sqrt{(2 + \sqrt{(2 + \sqrt{(2 + \sqrt{(2 - \sqrt{(2 - \sqrt{(2 - \sqrt{(2 + \sqrt{(2 - \sqrt{(2 - \sqrt{(2 + \sqrt{(2 - \sqrt{(2 + \sqrt{(2 + \sqrt{(2 + \sqrt{(2 + \sqrt{(2 - \sqrt{(2 - \sqrt{(2 - \sqrt{(2 + \sqrt{(2 - \sqrt{(2 + \sqrt{(2 + \sqrt{(2 + \sqrt{(2 + \sqrt{(2 - \sqrt{2}}/2$

For exact values, please see the table in Exact Values of integer angles.

Example of Sin 23.4°

Let us take an example having decimal like Sin 23.4°:

		Sign
90.00000000000000		
45.00000000000000	45.00000000000000	2
-22.50000000000000	22.50000000000000	-2
11.25000000000000	33.75000000000000	-2
-5.62500000000000	28.12500000000000	-2
-2.81250000000000	25.31250000000000	2
-1.40625000000000	23.90625000000000	2
-0.70312500000000	23.20312500000000	2
0.35156250000000	23.55468750000000	-2
-0.17578125000000	23.37890625000000	-2
……..	……..	….
32-steps	omitted	
……..	……..	….
0.00000000002046	23.39999999999920	2
0.00000000001023	23.40000000000940	2
-0.00000000000512	23.40000000000430	-2
-0.00000000000256	23.40000000000170	2
-0.00000000000128	23.40000000000050	2
-0.00000000000064	23.39999999999980	2
0.00000000000032	23.40000000000010	-2
-0.00000000000016	23.40000000000000	-2

Hence, Sin 23.4°= √(2 -√(2 -√(2 -√(2 + √(2 + √(2 + √(2 -√(2 -√(2 -√(2 -√(2 + √(2 + √(2 -√(2 + √(2 + √(2 + √(2 -√(2 -√(2 -√(2 -√(2 + √(2 + √(2 -√(2 + √(2 + √(2 + √(2 -√(2 -√(2 -√(2

31

$-\sqrt{(}2 + \sqrt{(}2 + \sqrt{(}2 -\sqrt{(}2 + \sqrt{(}2 + \sqrt{(}2 + \sqrt{(}2 -\sqrt{(}2 -\sqrt{(}2 -\sqrt{(}2 -\sqrt{(}2 + \sqrt{(}2 + \sqrt{(}2 -\sqrt{(}2 + \sqrt{(}2 + \sqrt{(}2 + \sqrt{(}2 -\sqrt{(}2 -\sqrt{2}\,]/2$

Cos 23.4° or for complementary angle, Sin 66.6°= $\sqrt{(}2 + \sqrt{(}2 -\sqrt{(}2 -\sqrt{(}2 + \sqrt{(}2 + \sqrt{(}2 + \sqrt{(}2 -\sqrt{(}2 -\sqrt{(}2 -\sqrt{(}2 + \sqrt{(}2 + \sqrt{(}2 -\sqrt{(}2 + \sqrt{(}2 + \sqrt{(}2 + \sqrt{(}2 -\sqrt{(}2 -\sqrt{(}2 -\sqrt{(}2 -\sqrt{(}2 + \sqrt{(}2 + \sqrt{(}2 -\sqrt{(}2 + \sqrt{(}2 + \sqrt{(}2 + \sqrt{(}2 -\sqrt{(}2 -\sqrt{(}2 -\sqrt{(}2 -\sqrt{(}2 + \sqrt{(}2 + \sqrt{(}2 -\sqrt{(}2 + \sqrt{(}2 + \sqrt{(}2 -\sqrt{(}2 -\sqrt{(}2 -\sqrt{(}2 -\sqrt{(}2 + \sqrt{(}2 + \sqrt{(}2 -\sqrt{(}2 + \sqrt{(}2 + \sqrt{(}2 -\sqrt{(}2 -\sqrt{2}\,]/2$

Exact Value of Sin 23.4°

Above values are accurate for 15-digits after decimal. In the accuracy chapter, we discussed there might be maximum error of ½ of $\sqrt[49]{2}$ in our Value of Sin 23.4°. To remove this error, we can convert this into exact values. For exact value, we carefully observe above result.

Sin 23.4°= $\sqrt{(}2 -\sqrt{(}2 -\sqrt{(}2 -\sqrt{(}2 + \sqrt{(}2 + \sqrt{(}2$
$\underline{+ \sqrt{(}2 -\sqrt{(}2 -\sqrt{(}2 -\sqrt{(}2 -\sqrt{(}2 + \sqrt{(}2 + \sqrt{(}2 -\sqrt{(}2 + \sqrt{(}2 + \sqrt{(}2}$
$+ \sqrt{(}2 -\sqrt{(}2 -\sqrt{(}2 -\sqrt{(}2 -\sqrt{(}2 + \sqrt{(}2 + \sqrt{(}2 -\sqrt{(}2 + \sqrt{(}2 + \sqrt{(}2$
$+ \sqrt{(}2 -\sqrt{(}2 -\sqrt{(}2 -\sqrt{(}2 -\sqrt{(}2 + \sqrt{(}2 + \sqrt{(}2 -\sqrt{(}2 + \sqrt{(}2 + \sqrt{(}2$
$+ \sqrt{(}2 -\sqrt{(}2 -\sqrt{(}2 -\sqrt{(}2 -\sqrt{(}2 + \sqrt{(}2 + \sqrt{(}2 -\sqrt{(}2 + \sqrt{(}2 + \sqrt{(}2$
$+ \sqrt{(}2 -\sqrt{(}2 -\sqrt{2}\,]/2$

So Exact value of Sin 23.4° is:

Sin 23.4°= Sin 23.4°= $\sqrt{(}2 -\sqrt{(}2 -\sqrt{(}2 -\sqrt{(}2 + \sqrt{(}2 + \sqrt{(}2$
$\underline{+ \sqrt{(}2 -\sqrt{(}2 -\sqrt{(}2 -\sqrt{(}2 -\sqrt{(}2 + \sqrt{(}2 + \sqrt{(}2 -\sqrt{(}2 + \sqrt{(}2 + \sqrt{(}2}\,]/2$

[Please care the sign of <u>repetition</u> has <u>underlined</u>.]

2.2 Sine of any angle

As discussed earlier and examples above, for an *appropriate accuracy*, this method gives *exact value of trigonometry ratios of any angle*, integer or decimal. In case our measurement of angle in grade or in radian, we should care the binary numbers only (not the process or result).

Following table has shown exact value of Sine of an angle from 1° to 45° (Cosine 89° - 45°) with appropriate accuracy of 15-digits after decimal. Need not to say, double of these will be chord (a) for double angle (2A). Changing the first '-ve' sign into '+ve', the result will be supplementary chord (b) for double angle (Crd. 2A). From 'a' and 'b' we can compute all of the sixteen trigonometric ratios given in sub-chapter 1.2. In the next sub-chapter, exact values for their angles have described.

A°	Sin A°
1	$\sqrt{(}2 - \sqrt{(}2 + \sqrt{(}2 + \sqrt{(}2 + \sqrt{(}2 + \sqrt{(}2 + \sqrt{(}2 - \sqrt{(}2 - \sqrt{(}2 - \sqrt{(}2 + \sqrt{(}2 - \sqrt{(}2 - \sqrt{(}2 + \sqrt{(}2 - \sqrt{(}2 + \sqrt{(}2 + \sqrt{(}2 + \sqrt{(}2 + \sqrt{(}2 - \sqrt{(}2 - \sqrt{(}2 - \sqrt{(}2 + \sqrt{(}2 - \sqrt{(}2 - \sqrt{(}2 + \sqrt{(}2 - \sqrt{(}2 + \sqrt{(}2 + \sqrt{(}2 + \sqrt{(}2 + \sqrt{(}2 - \sqrt{(}2 - \sqrt{(}2 - \sqrt{(}2 + \sqrt{(}2 - \sqrt{(}2 - \sqrt{(}2 + \sqrt{(}2 - \sqrt{(}2 + \sqrt{(}2 + \sqrt{(}2 + \sqrt{(}2 + \sqrt{(}2 - \sqrt{(}2 - \sqrt{(}2 - \sqrt{(}2 + \sqrt{(}2 - \sqrt{(}2 - \sqrt{2}]/2$
2	$\sqrt{(}2 - \sqrt{(}2 + \sqrt{(}2 + \sqrt{(}2 + \sqrt{(}2 + \sqrt{(}2 - \sqrt{(}2 - \sqrt{(}2 - \sqrt{(}2 + \sqrt{(}2 - \sqrt{(}2 - \sqrt{(}2 + \sqrt{(}2 - \sqrt{(}2 + \sqrt{(}2 + \sqrt{(}2 + \sqrt{(}2 + \sqrt{(}2 - \sqrt{(}2 - \sqrt{(}2 - \sqrt{(}2 + \sqrt{(}2 - \sqrt{(}2 - \sqrt{(}2 + \sqrt{(}2 - \sqrt{(}2 + \sqrt{(}2 - \sqrt{(}2 + \sqrt{(}2 + \sqrt{(}2 + \sqrt{(}2 - \sqrt{(}2 - \sqrt{(}2 + \sqrt{(}2 - \sqrt{(}2 - \sqrt{(}2 + \sqrt{(}2 - \sqrt{(}2 + \sqrt{(}2 + \sqrt{(}2 + \sqrt{(}2 - \sqrt{(}2 - \sqrt{(}2 - \sqrt{(}2 + \sqrt{(}2 - \sqrt{(}2 - \sqrt{(}2 + \sqrt{(}2 - \sqrt{(}2 + \sqrt{(}2 + \sqrt{(}2 + \sqrt{2}]/2$
3	$\sqrt{(}2 - \sqrt{(}2 + \sqrt{(}2 + \sqrt{(}2 + \sqrt{(}2 - \sqrt{(}2 - \sqrt{(}2 + \sqrt{(}2 + \sqrt{(}2 - \sqrt{(}2 - \sqrt{(}2 + \sqrt{(}2 + \sqrt{(}2 - \sqrt{(}2 - \sqrt{(}2 + \sqrt{(}2 + \sqrt{(}2 - \sqrt{(}2 - \sqrt{(}2 + \sqrt{(}2 + \sqrt{(}2 + \sqrt{(}2 - \sqrt{(}2 - \sqrt{(}2 + \sqrt{(}2 + \sqrt{(}2 - \sqrt{(}2 - \sqrt{(}2 + \sqrt{(}2 + \sqrt{(}2 - \sqrt{(}2 - \sqrt{(}2 + \sqrt{(}2 +$

$$\sqrt{(2-\sqrt{(2-\sqrt{(2+\sqrt{(2+\sqrt{(2-\sqrt{(2-\sqrt{(2+\sqrt{(2+\sqrt{(2-\sqrt{(2-\sqrt{(2+\sqrt{(2+\sqrt{(2-\sqrt{(2-\sqrt{(2+\sqrt{(2+\sqrt{2}}}}}}}}}}}}}}}}}}]/2$$

4. $$\sqrt{(2-\sqrt{(2+\sqrt{(2+\sqrt{(2+\sqrt{(2-\sqrt{(2-\sqrt{(2-\sqrt{(2+\sqrt{(2-\sqrt{(2-\sqrt{(2+\sqrt{(2+\sqrt{(2-\sqrt{(2+\sqrt{(2+\sqrt{(2+\sqrt{(2+\sqrt{(2-\sqrt{(2-\sqrt{(2-\sqrt{(2+\sqrt{(2-\sqrt{(2-\sqrt{(2+\sqrt{(2+\sqrt{(2+\sqrt{(2+\sqrt{(2+\sqrt{(2+\sqrt{(2-\sqrt{(2-\sqrt{(2-\sqrt{(2+\sqrt{(2-\sqrt{(2-\sqrt{(2+\sqrt{(2-\sqrt{(2+\sqrt{(2+\sqrt{(2+\sqrt{(2+\sqrt{(2-\sqrt{(2-\sqrt{(2-\sqrt{(2+\sqrt{(2-\sqrt{(2+\sqrt{(2-\sqrt{(2+\sqrt{(2+\sqrt{(2+\sqrt{(2+\sqrt{2}}]/2$$

5. $$\sqrt{(2-\sqrt{(2+\sqrt{(2+\sqrt{(2+\sqrt{(2-\sqrt{(2+\sqrt{(2+\sqrt{(2-\sqrt{(2+\sqrt{(2+\sqrt{(2-\sqrt{(2+\sqrt{(2+\sqrt{(2-\sqrt{(2+\sqrt{(2+\sqrt{(2-\sqrt{(2+\sqrt{(2+\sqrt{(2+\sqrt{(2+\sqrt{(2-\sqrt{(2+\sqrt{(2+\sqrt{(2-\sqrt{(2+\sqrt{(2+\sqrt{(2-\sqrt{(2+\sqrt{(2+\sqrt{(2+\sqrt{(2-\sqrt{(2+\sqrt{(2+\sqrt{(2-\sqrt{(2+\sqrt{(2+\sqrt{(2+\sqrt{(2-\sqrt{(2+\sqrt{(2+\sqrt{(2+\sqrt{(2-\sqrt{(2+\sqrt{(2+\sqrt{(2+\sqrt{(2-\sqrt{(2+\sqrt{2}}}]/2$$

6. $$\sqrt{(2-\sqrt{(2+\sqrt{(2+\sqrt{(2-\sqrt{(2-\sqrt{(2+\sqrt{(2+\sqrt{(2-\sqrt{(2-\sqrt{(2+\sqrt{(2+\sqrt{(2-\sqrt{(2-\sqrt{(2+\sqrt{(2+\sqrt{(2-\sqrt{(2-\sqrt{(2+\sqrt{(2+\sqrt{(2-\sqrt{(2-\sqrt{(2+\sqrt{(2+\sqrt{(2-\sqrt{(2-\sqrt{(2+\sqrt{(2+\sqrt{(2-\sqrt{(2-\sqrt{(2+\sqrt{(2+\sqrt{(2+\sqrt{(2-\sqrt{(2+\sqrt{(2+\sqrt{(2-\sqrt{(2-\sqrt{(2+\sqrt{(2+\sqrt{(2-\sqrt{(2-\sqrt{(2+\sqrt{(2+\sqrt{(2-\sqrt{(2-\sqrt{(2+\sqrt{(2+\sqrt{2}}]/2$$

7. $$\sqrt{(2-\sqrt{(2+\sqrt{(2+\sqrt{(2-\sqrt{(2-\sqrt{(2+\sqrt{(2-\sqrt{(2+\sqrt{(2+\sqrt{(2+\sqrt{(2+\sqrt{(2-\sqrt{(2-\sqrt{(2-\sqrt{(2+\sqrt{(2-\sqrt{(2-\sqrt{(2+\sqrt{(2-\sqrt{(2+\sqrt{(2+\sqrt{(2+\sqrt{(2+\sqrt{(2-\sqrt{(2-\sqrt{(2-\sqrt{(2+\sqrt{(2-\sqrt{(2-\sqrt{(2+\sqrt{(2-\sqrt{(2+\sqrt{(2+\sqrt{(2+\sqrt{(2+\sqrt{(2-\sqrt{(2-\sqrt{(2+\sqrt{(2-\sqrt{(2-\sqrt{(2+\sqrt{(2-\sqrt{(2+\sqrt{(2+\sqrt{(2+\sqrt{(2+\sqrt{(2-\sqrt{(2-\sqrt{(2-\sqrt{(2+\sqrt{(2-\sqrt{2}}]/2$$

8. $$\sqrt{(2-\sqrt{(2+\sqrt{(2+\sqrt{(2-\sqrt{(2-\sqrt{(2-\sqrt{(2+\sqrt{(2-\sqrt{(2-\sqrt{(2+\sqrt{(2-\sqrt{(2+\sqrt{(2+\sqrt{(2+\sqrt{(2+\sqrt{(2-\sqrt{(2-\sqrt{(2-\sqrt{(2+\sqrt{(2-\sqrt{(2-\sqrt{(2+\sqrt{(2-\sqrt{(2+\sqrt{(2+\sqrt{(2+\sqrt{(2+\sqrt{(2-\sqrt{(2-\sqrt{(2-\sqrt{(2+\sqrt{(2-\sqrt{(2-\sqrt{(2+\sqrt{(2-\sqrt{(2+\sqrt{(2+\sqrt{(2+\sqrt{(2+\sqrt{(2-\sqrt{(2-\sqrt{(2+\sqrt{(2+\sqrt{(2+\sqrt{(2-\sqrt{(2-\sqrt{(2-\sqrt{(2+\sqrt{(2-\sqrt{(2+\sqrt{(2-\sqrt{(2+\sqrt{(2+\sqrt{(2+\sqrt{(2+\sqrt{(2-\sqrt{2}}]/2$$

PRECISE-REWRITTEN METHOD

9 $\sqrt{(}2 - \sqrt{(}2 + \sqrt{(}2 + \sqrt{(}2 - \sqrt{2} \,]/2$

10 $\sqrt{(}2 - \sqrt{(}2 + \sqrt{(}2 + \sqrt{(}2 - \sqrt{(}2 + \sqrt{(}2 + \sqrt{(}2 - \sqrt{(}2 + \sqrt{(}2 + \sqrt{(}2 - \sqrt{(}2 + \sqrt{(}2 + \sqrt{(}2 - \sqrt{(}2 + \sqrt{(}2 + \sqrt{(}2 - \sqrt{(}2 + \sqrt{(}2 + \sqrt{(}2 - \sqrt{(}2 + \sqrt{(}2 + \sqrt{(}2 - \sqrt{(}2 + \sqrt{(}2 + \sqrt{(}2 - \sqrt{(}2 + \sqrt{(}2 + \sqrt{(}2 - \sqrt{(}2 + \sqrt{(}2 + \sqrt{(}2 - \sqrt{(}2 + \sqrt{(}2 + \sqrt{(}2 - \sqrt{(}2 + \sqrt{(}2 + \sqrt{(}2 - \sqrt{(}2 + \sqrt{(}2 + \sqrt{(}2 - \sqrt{(}2 + \sqrt{(}2 + \sqrt{(}2 - \sqrt{2} \,]/2$

11 $\sqrt{(}2 - \sqrt{(}2 + \sqrt{(}2 + \sqrt{(}2 - \sqrt{(}2 + \sqrt{(}2 + \sqrt{(}2 + \sqrt{(}2 + \sqrt{(}2 - \sqrt{(}2 - \sqrt{(}2 - \sqrt{(}2 + \sqrt{(}2 - \sqrt{(}2 - \sqrt{(}2 + \sqrt{(}2 - \sqrt{(}2 + \sqrt{(}2 + \sqrt{(}2 + \sqrt{(}2 + \sqrt{(}2 - \sqrt{(}2 - \sqrt{(}2 - \sqrt{(}2 + \sqrt{(}2 - \sqrt{(}2 - \sqrt{(}2 + \sqrt{(}2 - \sqrt{(}2 + \sqrt{(}2 + \sqrt{(}2 + \sqrt{(}2 - \sqrt{(}2 - \sqrt{(}2 - \sqrt{(}2 + \sqrt{(}2 - \sqrt{(}2 - \sqrt{(}2 + \sqrt{(}2 - \sqrt{(}2 + \sqrt{(}2 + \sqrt{(}2 + \sqrt{(}2 + \sqrt{(}2 - \sqrt{(}2 - \sqrt{(}2 - \sqrt{(}2 + \sqrt{2} \,]/2$

12 $\sqrt{(}2 - \sqrt{(}2 + \sqrt{(}2 - \sqrt{(}2 - \sqrt{(}2 + \sqrt{(}2 + \sqrt{(}2 - \sqrt{(}2 - \sqrt{(}2 + \sqrt{(}2 + \sqrt{(}2 - \sqrt{(}2 - \sqrt{(}2 + \sqrt{(}2 + \sqrt{(}2 - \sqrt{(}2 - \sqrt{(}2 + \sqrt{(}2 + \sqrt{(}2 - \sqrt{(}2 - \sqrt{(}2 + \sqrt{(}2 + \sqrt{(}2 - \sqrt{(}2 - \sqrt{(}2 + \sqrt{(}2 + \sqrt{(}2 - \sqrt{(}2 - \sqrt{(}2 + \sqrt{(}2 + \sqrt{(}2 - \sqrt{(}2 - \sqrt{(}2 + \sqrt{(}2 + \sqrt{(}2 - \sqrt{(}2 - \sqrt{(}2 + \sqrt{(}2 + \sqrt{(}2 - \sqrt{(}2 - \sqrt{(}2 + \sqrt{(}2 + \sqrt{2} \,]/2$

13 $\sqrt{(}2 - \sqrt{(}2 + \sqrt{(}2 - \sqrt{(}2 - \sqrt{(}2 + \sqrt{(}2 - \sqrt{(}2 - \sqrt{(}2 + \sqrt{(}2 - \sqrt{(}2 + \sqrt{(}2 + \sqrt{(}2 + \sqrt{(}2 + \sqrt{(}2 - \sqrt{(}2 - \sqrt{(}2 - \sqrt{(}2 + \sqrt{(}2 - \sqrt{(}2 - \sqrt{(}2 + \sqrt{(}2 - \sqrt{(}2 + \sqrt{(}2 + \sqrt{(}2 + \sqrt{(}2 - \sqrt{(}2 - \sqrt{(}2 - \sqrt{(}2 + \sqrt{(}2 - \sqrt{(}2 + \sqrt{(}2 - \sqrt{(}2 + \sqrt{(}2 + \sqrt{(}2 + \sqrt{(}2 + \sqrt{(}2 - \sqrt{(}2 - \sqrt{(}2 - \sqrt{(}2 - \sqrt{(}2 + \sqrt{(}2 - \sqrt{(}2 + \sqrt{(}2 - \sqrt{(}2 + \sqrt{(}2 + \sqrt{(}2 + \sqrt{2} \,]/2$

14 $\sqrt{(}2 - \sqrt{(}2 + \sqrt{(}2 - \sqrt{(}2 - \sqrt{(}2 + \sqrt{(}2 - \sqrt{(}2 + \sqrt{(}2 + \sqrt{(}2 + \sqrt{(}2 + \sqrt{(}2 - \sqrt{(}2 - \sqrt{(}2 - \sqrt{(}2 + \sqrt{(}2 - \sqrt{(}2 - \sqrt{(}2 + \sqrt{(}2 - \sqrt{(}2 + \sqrt{(}2 + \sqrt{(}2 + \sqrt{(}2 + \sqrt{(}2 - \sqrt{(}2 - \sqrt{(}2 - \sqrt{(}2 + \sqrt{(}2 - \sqrt{(}2 - \sqrt{(}2 + \sqrt{(}2 - \sqrt{(}2 + \sqrt{(}2 + \sqrt{(}2 + \sqrt{(}2 + \sqrt{(}2 - \sqrt{(}2 - \sqrt{(}2 - $

√(2 - √(2 + √(2 - √(2 - √(2 + √(2 - √(2 + √(2 + √(2 + √(2 + √(2 - √2]/2

15 √(2 - √(2 + √(2 - √(2 - √(2 - √(2 - √(2 - √(2 - √(2 - √(2
    - √(2 - √(2 - √(2 - √(2 - √(2 - √(2 - √(2 - √(2 - √(2 -
    √(2 - √(2 - √(2 - √(2 - √(2 - √(2 - √(2 - √(2 - √(2 - √(2
    - √(2 - √(2 - √(2 - √(2 - √(2 - √(2 - √(2 - √(2 - √(2 -
    √(2 - √(2 - √(2 - √(2 - √(2 - √(2 - √(2 - √(2 - √(2 - √(2
    - √(2 - √2]/2

16 √(2 - √(2 + √(2 - √(2 - √(2 - √(2 + √(2 - √(2 - √(2 +
    √(2 - √(2 + √(2 + √(2 + √(2 + √(2 - √(2 - √(2 - √(2 +
    √(2 - √(2 - √(2 + √(2 - √(2 + √(2 + √(2 + √(2 + √(2 -
    √(2 - √(2 - √(2 + √(2 - √(2 - √(2 + √(2 - √(2 + √(2 +
    √(2 + √(2 + √(2 - √(2 - √(2 - √(2 + √(2 - √(2 - √(2 +
    √(2 - √(2 + √(2 + √(2 + √2]/2

17 √(2 - √(2 + √(2 - √(2 + √(2 - √(2 + √(2 + √(2 + √(2
    + √(2 - √(2 - √(2 - √(2 + √(2 - √(2 - √(2 + √(2 - √(2 +
    √(2 + √(2 + √(2 + √(2 - √(2 - √(2 - √(2 + √(2 - √(2 -
    √(2 + √(2 - √(2 + √(2 + √(2 + √(2 + √(2 - √(2 - √(2 -
    √(2 + √(2 - √(2 - √(2 + √(2 - √(2 + √(2 + √(2 + √(2
    + √(2 - √(2 - √(2 - √(2 + √2]/2

18 √(2 - √(2 + √(2 - √(2 + √(2 - √(2 + √(2 - √(2 + √(2 -
    √(2 + √(2 - √(2 + √(2 - √(2 + √(2 - √(2 + √(2 - √(2 +
    √(2 - √(2 + √(2 - √(2 + √(2 - √(2 + √(2 - √(2 + √(2 -
    √(2 + √(2 - √(2 + √(2 - √(2 + √(2 - √(2 + √(2 - √(2 +
    √(2 - √(2 + √(2 - √(2 + √(2 - √(2 + √(2 - √(2 + √(2 -
    √(2 + √(2 - √(2 + √(2 - √2]/2

19 √(2 - √(2 + √(2 - √(2 + √(2 - √(2 - √(2 + √(2 - √(2 +
    √(2 + √(2 + √(2 + √(2 - √(2 - √(2 - √(2 + √(2 - √(2 -
    √(2 + √(2 - √(2 + √(2 + √(2 + √(2 + √(2 - √(2 - √(2 -
    √(2 + √(2 - √(2 - √(2 + √(2 - √(2 + √(2 + √(2 + √(2
    + √(2 - √(2 - √(2 - √(2 + √(2 - √(2 - √(2 + √(2 - √(2 +
    √(2 + √(2 + √(2 + √(2 - √2]/2

20 √(2 - √(2 + √(2 - √(2 + √(2 + √(2 - √(2 + √(2 + √(2 -
    √(2 + √(2 + √(2 - √(2 + √(2 + √(2 - √(2 + √(2 + √(2

PRECISE-REWRITTEN METHOD

$-\sqrt{(2 + \sqrt{(2 + \sqrt{(2 - \sqrt{(2 + \sqrt{(2 + \sqrt{(2 - \sqrt{(2 + \sqrt{(2 + }}}}}}}}$
$\sqrt{(2 - \sqrt{(2 + \sqrt{(2 + \sqrt{(2 - \sqrt{(2 + \sqrt{(2 + \sqrt{(2 - \sqrt{(2 + \sqrt{(2 }}}}}}}}}}$
$+ \sqrt{(2 - \sqrt{(2 + \sqrt{(2 + \sqrt{(2 - \sqrt{(2 + \sqrt{(2 + \sqrt{(2 - \sqrt{(2 + }}}}}}}}$
$\sqrt{(2 + \sqrt{(2 - \sqrt{(2 + \sqrt{(2 + \sqrt{(2 - \sqrt{(2 + \sqrt{2} }}}}}}]/2$

21   $\sqrt{(2 - \sqrt{(2 + \sqrt{(2 - \sqrt{(2 + \sqrt{(2 + \sqrt{(2 - \sqrt{(2 - \sqrt{(2 + \sqrt{(2 + }}}}}}}}}$
$\sqrt{(2 - \sqrt{(2 - \sqrt{(2 + \sqrt{(2 + \sqrt{(2 - \sqrt{(2 - \sqrt{(2 + \sqrt{(2 + \sqrt{(2 - }}}}}}}}}}$
$\sqrt{(2 - \sqrt{(2 + \sqrt{(2 + \sqrt{(2 - \sqrt{(2 - \sqrt{(2 + \sqrt{(2 + \sqrt{(2 - \sqrt{(2 - }}}}}}}}}}$
$\sqrt{(2 + \sqrt{(2 + \sqrt{(2 - \sqrt{(2 - \sqrt{(2 + \sqrt{(2 + \sqrt{(2 - \sqrt{(2 - \sqrt{(2 + }}}}}}}}}}$
$\sqrt{(2 + \sqrt{(2 - \sqrt{(2 - \sqrt{(2 + \sqrt{(2 + \sqrt{(2 - \sqrt{(2 - \sqrt{(2 + \sqrt{(2 + }}}}}}}}}}$
$\sqrt{(2 - \sqrt{(2 - \sqrt{(2 + \sqrt{(2 + \sqrt{2} }}}}]/2$

22   $\sqrt{(2 - \sqrt{(2 + \sqrt{(2 - \sqrt{(2 + \sqrt{(2 + \sqrt{(2 + \sqrt{(2 + \sqrt{(2 - \sqrt{(2 - }}}}}}}}}$
$\sqrt{(2 - \sqrt{(2 + \sqrt{(2 - \sqrt{(2 - \sqrt{(2 + \sqrt{(2 - \sqrt{(2 + \sqrt{(2 + \sqrt{(2 + }}}}}}}}}}$
$\sqrt{(2 + \sqrt{(2 - \sqrt{(2 - \sqrt{(2 - \sqrt{(2 + \sqrt{(2 - \sqrt{(2 - \sqrt{(2 + \sqrt{(2 - }}}}}}}}}}$
$\sqrt{(2 + \sqrt{(2 + \sqrt{(2 + \sqrt{(2 + \sqrt{(2 - \sqrt{(2 - \sqrt{(2 - \sqrt{(2 + \sqrt{(2 - }}}}}}}}}}$
$\sqrt{(2 - \sqrt{(2 + \sqrt{(2 - \sqrt{(2 + \sqrt{(2 + \sqrt{(2 + \sqrt{(2 + \sqrt{(2 - \sqrt{(2 - }}}}}}}}}}$
$\sqrt{(2 - \sqrt{(2 + \sqrt{(2 - \sqrt{(2 - \sqrt{2} }}}}]/2$

23   $\sqrt{(2 - \sqrt{(2 - \sqrt{(2 - \sqrt{(2 + \sqrt{(2 + \sqrt{(2 + \sqrt{(2 + \sqrt{(2 - \sqrt{(2 - }}}}}}}}}$
$\sqrt{(2 - \sqrt{(2 + \sqrt{(2 - \sqrt{(2 - \sqrt{(2 + \sqrt{(2 - \sqrt{(2 + \sqrt{(2 + \sqrt{(2 + }}}}}}}}}}$
$\sqrt{(2 + \sqrt{(2 - \sqrt{(2 - \sqrt{(2 - \sqrt{(2 + \sqrt{(2 - \sqrt{(2 - \sqrt{(2 + \sqrt{(2 - }}}}}}}}}}$
$\sqrt{(2 + \sqrt{(2 + \sqrt{(2 + \sqrt{(2 + \sqrt{(2 - \sqrt{(2 - \sqrt{(2 - \sqrt{(2 + \sqrt{(2 - }}}}}}}}}}$
$\sqrt{(2 - \sqrt{(2 + \sqrt{(2 - \sqrt{(2 + \sqrt{(2 + \sqrt{(2 + \sqrt{(2 + \sqrt{(2 - \sqrt{(2 - }}}}}}}}}}$
$\sqrt{(2 - \sqrt{(2 + \sqrt{(2 - \sqrt{(2 - \sqrt{2} }}}}]/2$

24   $\sqrt{(2 - \sqrt{(2 - \sqrt{(2 - \sqrt{(2 + \sqrt{(2 + \sqrt{(2 - \sqrt{(2 - \sqrt{(2 + \sqrt{(2 + }}}}}}}}}$
$\sqrt{(2 - \sqrt{(2 - \sqrt{(2 + \sqrt{(2 + \sqrt{(2 - \sqrt{(2 - \sqrt{(2 + \sqrt{(2 + \sqrt{(2 - }}}}}}}}}}$
$\sqrt{(2 - \sqrt{(2 + \sqrt{(2 + \sqrt{(2 - \sqrt{(2 - \sqrt{(2 + \sqrt{(2 + \sqrt{(2 - \sqrt{(2 - }}}}}}}}}}$
$\sqrt{(2 + \sqrt{(2 + \sqrt{(2 - \sqrt{(2 - \sqrt{(2 + \sqrt{(2 + \sqrt{(2 - \sqrt{(2 - \sqrt{(2 + }}}}}}}}}}$
$\sqrt{(2 + \sqrt{(2 - \sqrt{(2 - \sqrt{(2 + \sqrt{(2 + \sqrt{(2 - \sqrt{(2 - \sqrt{(2 + \sqrt{(2 + }}}}}}}}}}$
$\sqrt{(2 - \sqrt{(2 - \sqrt{(2 + \sqrt{(2 + \sqrt{2} }}}}]/2$

25   $\sqrt{(2 - \sqrt{(2 - \sqrt{(2 - \sqrt{(2 + \sqrt{(2 + \sqrt{(2 - \sqrt{(2 + \sqrt{(2 + \sqrt{(2 - }}}}}}}}}$
$\sqrt{(2 + \sqrt{(2 + \sqrt{(2 - \sqrt{(2 + \sqrt{(2 + \sqrt{(2 - \sqrt{(2 + \sqrt{(2 + \sqrt{(2 }}}}}}}}}}$
$- \sqrt{(2 + \sqrt{(2 + \sqrt{(2 - \sqrt{(2 + \sqrt{(2 + \sqrt{(2 - \sqrt{(2 + \sqrt{(2 + }}}}}}}}$
$\sqrt{(2 - \sqrt{(2 + \sqrt{(2 + \sqrt{(2 - \sqrt{(2 + \sqrt{(2 + \sqrt{(2 - \sqrt{(2 + \sqrt{(2 }}}}}}}}}$
$+ \sqrt{(2 - \sqrt{(2 + \sqrt{(2 + \sqrt{(2 - \sqrt{(2 + \sqrt{(2 + \sqrt{(2 - \sqrt{(2 + }}}}}}}}$
$\sqrt{(2 + \sqrt{(2 - \sqrt{(2 + \sqrt{(2 + \sqrt{(2 - \sqrt{(2 + \sqrt{2} }}}}}}]/2$

26. $\sqrt{(2 - \sqrt{(2 - \sqrt{(2 - \sqrt{(2 + \sqrt{(2 - \sqrt{(2 - \sqrt{(2 + \sqrt{(2 - \sqrt{(2 + \sqrt{(2 + \sqrt{(2 + \sqrt{(2 - \sqrt{(2 - \sqrt{(2 - \sqrt{(2 + \sqrt{(2 - \sqrt{(2 - \sqrt{(2 + \sqrt{(2 + \sqrt{(2 - \sqrt{(2 + \sqrt{(2 + \sqrt{(2 + \sqrt{(2 + \sqrt{(2 - \sqrt{(2 - \sqrt{(2 + \sqrt{(2 - \sqrt{(2 - \sqrt{(2 + \sqrt{(2 - \sqrt{(2 + \sqrt{(2 + \sqrt{(2 + \sqrt{(2 - \sqrt{(2 - \sqrt{(2 - \sqrt{(2 + \sqrt{(2 - \sqrt{(2 - \sqrt{(2 + \sqrt{(2 - \sqrt{(2 + \sqrt{(2 + \sqrt{(2 + \sqrt{(2 - \sqrt{2}\,]}}/2$

27. $\sqrt{(2 - \sqrt{(2 - \sqrt{(2 - \sqrt{(2 + \sqrt{(2 - \sqrt{(2 + \sqrt{(2 - \sqrt{(2 + \sqrt{(2 - \sqrt{(2 + \sqrt{(2 - \sqrt{(2 + \sqrt{(2 - \sqrt{(2 + \sqrt{(2 - \sqrt{(2 + \sqrt{(2 - \sqrt{(2 + \sqrt{(2 + \sqrt{(2 - \sqrt{(2 + \sqrt{(2 - \sqrt{(2 + \sqrt{(2 - \sqrt{(2 + \sqrt{(2 - \sqrt{(2 + \sqrt{(2 + \sqrt{(2 - \sqrt{(2 + \sqrt{(2 - \sqrt{(2 + \sqrt{(2 + \sqrt{(2 + \sqrt{(2 - \sqrt{(2 + \sqrt{(2 - \sqrt{(2 + \sqrt{(2 - \sqrt{(2 + \sqrt{(2 - \sqrt{(2 + \sqrt{(2 - \sqrt{(2 + \sqrt{(2 - \sqrt{2}\,]}}/2$

28. $\sqrt{(2 - \sqrt{(2 - \sqrt{(2 - \sqrt{(2 + \sqrt{(2 - \sqrt{(2 + \sqrt{(2 + \sqrt{(2 + \sqrt{(2 + \sqrt{(2 - \sqrt{(2 - \sqrt{(2 - \sqrt{(2 + \sqrt{(2 - \sqrt{(2 - \sqrt{(2 + \sqrt{(2 - \sqrt{(2 + \sqrt{(2 + \sqrt{(2 + \sqrt{(2 - \sqrt{(2 - \sqrt{(2 - \sqrt{(2 + \sqrt{(2 - \sqrt{(2 - \sqrt{(2 + \sqrt{(2 + \sqrt{(2 + \sqrt{(2 + \sqrt{(2 - \sqrt{(2 - \sqrt{(2 + \sqrt{(2 - \sqrt{(2 + \sqrt{(2 + \sqrt{(2 + \sqrt{(2 - \sqrt{(2 + \sqrt{(2 - \sqrt{(2 - \sqrt{(2 + \sqrt{2}\,]}}}/2$

29. $\sqrt{(2 - \sqrt{(2 - \sqrt{(2 - \sqrt{(2 - \sqrt{(2 - \sqrt{(2 + \sqrt{(2 - \sqrt{(2 - \sqrt{(2 + \sqrt{(2 - \sqrt{(2 + \sqrt{(2 + \sqrt{(2 + \sqrt{(2 + \sqrt{(2 - \sqrt{(2 - \sqrt{(2 - \sqrt{(2 + \sqrt{(2 - \sqrt{(2 - \sqrt{(2 + \sqrt{(2 - \sqrt{(2 + \sqrt{(2 + \sqrt{(2 + \sqrt{(2 - \sqrt{(2 - \sqrt{(2 - \sqrt{(2 + \sqrt{(2 - \sqrt{(2 + \sqrt{(2 - \sqrt{(2 + \sqrt{(2 - \sqrt{(2 + \sqrt{(2 + \sqrt{(2 + \sqrt{(2 + \sqrt{(2 - \sqrt{(2 - \sqrt{(2 - \sqrt{(2 + \sqrt{(2 - \sqrt{(2 - \sqrt{(2 + \sqrt{(2 - \sqrt{(2 + \sqrt{(2 + \sqrt{(2 + \sqrt{2}\,]}}}/2$

30. $\sqrt{(2 - \sqrt{(2 \quad \sqrt{2}\,]}}/2$

31. $\sqrt{(2 - \sqrt{(2 - \sqrt{(2 - \sqrt{(2 - \sqrt{(2 + \sqrt{(2 - \sqrt{(2 + \sqrt{(2 + \sqrt{(2 + \sqrt{(2 + \sqrt{(2 - \sqrt{(2 - \sqrt{(2 - \sqrt{(2 + \sqrt{(2 - \sqrt{(2 - \sqrt{(2 + \sqrt{(2 - \sqrt{(2 + \sqrt{(2 + \sqrt{(2 + \sqrt{(2 + \sqrt{(2 - \sqrt{(2 - \sqrt{(2 - \sqrt{(2 + \sqrt{(2 - \sqrt{(2 + \sqrt{(2 - \sqrt{(2 + \sqrt{(2 - \sqrt{(2 + \sqrt{(2 + \sqrt{(2 + \sqrt{(2 + \sqrt{(2 - \sqrt{(2 - \sqrt{(2 -}}$

√(2 - √(2 + √(2 - √(2 - √(2 + √(2 - √(2 + √(2 + √(2 + √(2 + √(2 - √2]/2

32   √(2 - √(2 - √(2 - √(2 - √(2 + √(2 - √(2 - √(2 + √(2 - √(2 + √(2 + √(2 + √(2 + √(2 - √(2 - √(2 + √(2 - √(2 - √(2 + √(2 - √(2 + √(2 + √(2 + √(2 - √(2 - √(2 - √(2 + √(2 - √(2 - √(2 + √(2 + √(2 + √(2 + √(2 - √(2 - √(2 - √(2 - √(2 + √(2 - √(2 - √(2 + √(2 - √(2 + √(2 + √(2 + √2]/2

33   √(2 - √(2 - √(2 - √(2 - √(2 + √(2 + √(2 - √(2 - √(2 + √(2 + √(2 - √(2 - √(2 + √(2 + √(2 - √(2 - √(2 + √(2 + √(2 - √(2 - √(2 - √(2 - √(2 + √(2 + √(2 - √(2 - √(2 + √(2 + √(2 - √(2 - √(2 + √(2 + √(2 - √(2 - √(2 + √(2 + √(2 - √(2 - √(2 - √(2 + √(2 + √(2 - √(2 - √(2 + √(2 + √2]/2

34   +√(2 - √(2 - √(2 + √(2 - √(2 + √(2 + √(2 + √(2 + √(2 - √(2 - √(2 - √(2 + √(2 - √(2 - √(2 + √(2 - √(2 + √(2 + √(2 + √(2 - √(2 - √(2 - √(2 + √(2 - √(2 - √(2 + √(2 - √(2 + √(2 + √(2 + √(2 - √(2 - √(2 - √(2 + √(2 - √(2 - √(2 + √(2 - √(2 + √(2 + √(2 + √(2 + √(2 - √(2 - √(2 - √(2 + √2]/2

35   √(2 - √(2 - √(2 + √(2 - √(2 + √(2 + √(2 - √(2 + √(2 + √(2 - √(2 + √(2 + √(2 - √(2 + √(2 + √(2 - √(2 + √(2 + √(2 - √(2 + √(2 + √(2 - √(2 + √(2 + √(2 - √(2 + √(2 + √(2 - √(2 + √(2 + √(2 - √(2 + √(2 + √(2 - √(2 + √(2 + √(2 - √(2 + √(2 + √(2 - √(2 + √2]/2

36   √(2 - √(2 - √(2 + √(2 - √(2 + √(2 - √(2 + √(2 - √(2 + √(2 - √(2 + √(2 - √(2 + √(2 - √(2 + √(2 - √(2 + √(2 - √(2 + √(2 - √(2 + √(2 - √(2 + √(2 - √(2 + √(2 - √(2 - √(2 + √(2 - √(2 + √(2 - √(2 + √(2 - √(2 + √(2 - √(2 + √(2 - √(2 + √(2 - √(2 + √(2 - √(2 + √(2 - √(2 + √(2 - √(2 + √(2 - √2]/2

37. $\sqrt{(2 - \sqrt{(2 - \sqrt{(2 + \sqrt{(2 - \sqrt{(2 - \sqrt{(2 - \sqrt{(2 + \sqrt{(2 - \sqrt{(2 - \sqrt{(2 + \sqrt{(2 - \sqrt{(2 + \sqrt{(2 + \sqrt{(2 + \sqrt{(2 - \sqrt{(2 - \sqrt{(2 - \sqrt{(2 + \sqrt{(2 - \sqrt{(2 - \sqrt{(2 + \sqrt{(2 - \sqrt{(2 + \sqrt{(2 + \sqrt{(2 + \sqrt{(2 + \sqrt{(2 - \sqrt{(2 - \sqrt{(2 - \sqrt{(2 + \sqrt{(2 - \sqrt{(2 - \sqrt{(2 + \sqrt{(2 - \sqrt{(2 + \sqrt{(2 + \sqrt{(2 + \sqrt{(2 + \sqrt{(2 - \sqrt{(2 - \sqrt{(2 - \sqrt{(2 + \sqrt{(2 - \sqrt{(2 - \sqrt{2}}]/2$

38. $\sqrt{(2 - \sqrt{(2 - \sqrt{(2 + \sqrt{(2 - \sqrt{(2 - \sqrt{(2 + \sqrt{(2 - \sqrt{(2 + \sqrt{(2 + \sqrt{(2 + \sqrt{(2 + \sqrt{(2 - \sqrt{(2 - \sqrt{(2 - \sqrt{(2 + \sqrt{(2 - \sqrt{(2 - \sqrt{(2 + \sqrt{(2 - \sqrt{(2 + \sqrt{(2 + \sqrt{(2 + \sqrt{(2 + \sqrt{(2 - \sqrt{(2 - \sqrt{(2 - \sqrt{(2 + \sqrt{(2 - \sqrt{(2 - \sqrt{(2 + \sqrt{(2 - \sqrt{(2 + \sqrt{(2 + \sqrt{(2 + \sqrt{(2 - \sqrt{(2 - \sqrt{(2 - \sqrt{(2 + \sqrt{(2 - \sqrt{(2 - \sqrt{(2 + \sqrt{(2 - \sqrt{(2 + \sqrt{(2 + \sqrt{(2 + \sqrt{(2 + \sqrt{(2 - \sqrt{2}}]/2$

39. $\sqrt{(2 - \sqrt{(2 - \sqrt{(2 + \sqrt{(2 - \sqrt{(2 - \sqrt{(2 + \sqrt{(2 + \sqrt{(2 - \sqrt{(2 - \sqrt{(2 + \sqrt{(2 + \sqrt{(2 - \sqrt{(2 - \sqrt{(2 + \sqrt{(2 + \sqrt{(2 - \sqrt{(2 - \sqrt{(2 + \sqrt{(2 + \sqrt{(2 - \sqrt{(2 - \sqrt{(2 + \sqrt{(2 + \sqrt{(2 - \sqrt{(2 - \sqrt{(2 - \sqrt{(2 + \sqrt{(2 + \sqrt{(2 - \sqrt{(2 - \sqrt{(2 + \sqrt{(2 + \sqrt{(2 - \sqrt{(2 - \sqrt{(2 + \sqrt{(2 + \sqrt{(2 - \sqrt{(2 - \sqrt{(2 + \sqrt{(2 + \sqrt{(2 + \sqrt{(2 - \sqrt{(2 - \sqrt{(2 + \sqrt{(2 + \sqrt{2}}]/2$

40. $\sqrt{(2 - \sqrt{(2 - \sqrt{(2 + \sqrt{(2 + \sqrt{(2 - \sqrt{(2 + \sqrt{(2 + \sqrt{(2 - \sqrt{(2 + \sqrt{(2 + \sqrt{(2 - \sqrt{(2 + \sqrt{(2 + \sqrt{(2 - \sqrt{(2 + \sqrt{(2 + \sqrt{(2 - \sqrt{(2 + \sqrt{(2 + \sqrt{(2 - \sqrt{(2 + \sqrt{(2 + \sqrt{(2 - \sqrt{(2 + \sqrt{(2 + \sqrt{(2 - \sqrt{(2 + \sqrt{(2 + \sqrt{(2 + \sqrt{(2 - \sqrt{(2 + \sqrt{(2 + \sqrt{(2 - \sqrt{(2 + \sqrt{(2 + \sqrt{(2 - \sqrt{(2 + \sqrt{(2 + \sqrt{(2 - \sqrt{(2 + \sqrt{(2 + \sqrt{(2 - \sqrt{(2 + \sqrt{2}}}]/2$

41. $\sqrt{(2 - \sqrt{(2 - \sqrt{(2 + \sqrt{(2 + \sqrt{(2 - \sqrt{(2 - \sqrt{(2 - \sqrt{(2 + \sqrt{(2 - \sqrt{(2 - \sqrt{(2 + \sqrt{(2 - \sqrt{(2 + \sqrt{(2 + \sqrt{(2 + \sqrt{(2 - \sqrt{(2 - \sqrt{(2 - \sqrt{(2 + \sqrt{(2 - \sqrt{(2 + \sqrt{(2 + \sqrt{(2 + \sqrt{(2 + \sqrt{(2 - \sqrt{(2 - \sqrt{(2 - \sqrt{(2 + \sqrt{(2 - \sqrt{(2 - \sqrt{(2 + \sqrt{(2 - \sqrt{(2 + \sqrt{(2 + \sqrt{(2 + \sqrt{(2 + \sqrt{(2 - \sqrt{(2 - \sqrt{(2 - \sqrt{(2 + \sqrt{(2 - \sqrt{(2\ \sqrt{2}}}]/2$

42. $\sqrt{(2 - \sqrt{(2 - \sqrt{(2 + \sqrt{(2 + \sqrt{(2 - \sqrt{(2 - \sqrt{(2 + \sqrt{(2 + \sqrt{(2 - \sqrt{(2 - \sqrt{(2 + \sqrt{(2 + \sqrt{(2 - \sqrt{(2 - \sqrt{(2 + \sqrt{(2 + \sqrt{(2 - \sqrt{(2 - \sqrt{(2 + \sqrt{(2 + \sqrt{(2 - \sqrt{(2 - \sqrt{(2 + \sqrt{(2 + \sqrt{(2 - \sqrt{(2 - \sqrt{(2 + \sqrt{(2 + \sqrt{(2 - \sqrt{(2 - \sqrt{(2 + \sqrt{(2 + \sqrt{(2 - \sqrt{(2 - \sqrt{(2 + \sqrt{(2 +

PRECISE-REWRITTEN METHOD

√(2 - √(2 - √(2 + √(2 + √(2 - √(2 - √(2 + √(2 + √(2 - √(2 - √(2 + √(2 + √2]/2

43 √(2 - √(2 - √(2 + √(2 + √(2 + √(2 - √(2 - √(2 + √(2 - √(2 - √(2 + √(2 - √(2 + √(2 + √(2 + √(2 + √(2 - √(2 - √(2 - √(2 + √(2 - √(2 - √(2 + √(2 - √(2 + √(2 + √(2 + √(2 + √(2 - √(2 - √(2 - √(2 + √(2 - √(2 - √(2 + √(2 - √(2 + √(2 + √(2 + √(2 + √(2 - √(2 - √(2 - √(2 + √(2 - √(2 - √2]/2

44 √(2 - √(2 - √(2 + √(2 + √(2 + √(2 - √(2 - √(2 - √(2 + √(2 - √(2 - √(2 + √(2 - √(2 + √(2 + √(2 + √(2 - √(2 - √(2 - √(2 + √(2 - √(2 - √(2 + √(2 - √(2 + √(2 + √(2 + √(2 - √(2 - √(2 - √(2 + √(2 - √(2 + √(2 - √(2 + √(2 + √(2 + √(2 + √(2 - √(2 - √(2 - √(2 + √(2 - √(2 - √2]/2

45 √2/2

2.3 Exact Values of integer angles

If we carefully observe above values of Sine of integer angles, they are in *repeating pattern* of infinitive occurrence. For an integer angles, they are not only on repeating pattern, but also the pattern itself has *rule of repetition*. For exact values in trigonometry for integer angles, (see below for all regular polygon), following *rule of repetition* applies:

> All integer angles (and regular polygons) has exact value in repetitive pattern, if computed using Precise-Rewritten method.
>
> First *three radicals* of odd integer angles (1°, 3°, 5°, ...) are non-repetitive, remaining are in repetitive pattern.
>
> First *two radicals* of integer angles which are divisible by 4 (4°, 8°, 12°, ...) are non-repetitive, remaining are in repetitive pattern.
>
> First *one radical* of even integer angles except divisible by 4 (2°, 6°, 10°, ...) are non-repetitive, remaining are in repetitive pattern.
>
> Multiples of 3 has good and easy alternative. Even angles have alternative of 2-radicals or 1- radical.

Here is the list of exact value of Sine of integer angles (A). Cosine of complementary angle (90°-A) is same as Sin A. the list contains angles up to 45°, therefore the second sign is '-ve'. Converting this first sign into '+', that is the Sine of complementary angle (90°-A). Value without division by 2 will be the chord (a) of double angle in all cases.

Sin 1° (Crd. 2°)

This is odd integer angle. First three radicals will be non-repetitive. Then after there will be repeating pattern of nested radicals as:

Sin 1°= √(2 -√(2 +√(2

+√(2 +√(2 +√(2 +√(2 -√(2 -√(2 -√(2 +√(2 -√(2 -√(2 +√(2 -√(2

PRECISE-REWRITTEN METHOD

+√(2 +√(2 +√(2 +√(2 -√(2 -√(2 -√(2 +√(2 -√(2 -√(2 +√(2 -√(2
+√(2 +√(2 +√(2 +√(2 -√(2 -√(2 -√(2 +√(2 -√(2 -√(2 +√(2 -√(2
+√(2 +√(2 +√(2 +√(2 -√(2 -√(2 -√(2 +√(2 -√(2 -√(2 +√(2 -√(2
+√(2 +√(2 +√(2]/2

Sin 1° = ½ √(2 -√(2 +√(2

<u>+√(2 +√(2 +√(2 +√(2 -√(2 -√(2 -√(2 +√(2 -√(2 -√(2 +√(2 -√(2</u>
]

Sin 89° = ½ √(2 + √(2 +√(2

<u>+√(2 +√(2 +√(2 +√(2 -√(2 -√(2 -√(2 +√(2 -√(2 -√(2 +√(2 -√(2</u>]

[Please care the sign of <u>repetition</u> has <u>underlined</u>.]

Sin 2° (Crd. 4°)

This is even integer angle (and not divisible by 4). Only the first radical will be non-repetitive. Then after there will be repeating pattern of nested radicals as:

Sin 2° = √(2

<u>-√(2 +√(2 +√(2 +√(2 +√(2 -√(2 -√(2 -√(2 +√(2 -√(2 -√(2 +√(2</u>
-√(2 +√(2 +√(2 +√(2 +√(2 -√(2 -√(2 -√(2 +√(2 -√(2 -√(2 +√(2
-√(2 + √(2 +√(2 +√(2 +√(2 -√(2 -√(2 -√(2 +√(2 -√(2 -√(2 +√(2
-√(2 + √(2 +√(2 +√(2 +√(2 -√(2 -√(2 -√(2 +√(2 -√(2 -√(2 +√(2
-√(2 + √(2 +√(2 +√(2]/2

Therefore,

Sin 2° = ½ √(2 <u>-√(2 +√(2 +√(2 +√(2 +√(2 -√(2 -√(2 -√(2 +√(2</u>
-√(2 -√(2 +√(2]

Complementary angle of 2° is 88°. This is complementary of 2° as well as perfectly divisible by 4. Therefore, it will have two alternative solutions with same value:

First alternative

Sin 88° = ½ √(2 − <u>-√(2 +√(2 +√(2 +√(2 +√(2 -√(2 -√(2 -√(2</u>
+√(2 -√(2 -√(2 +√(2]

Second alternative

Sin 88° = √(2 + √(2

43

$+\sqrt{(2} +\sqrt{(2} +\sqrt{(2} +\sqrt{(2} -\sqrt{(2} -\sqrt{(2} -\sqrt{(2} +\sqrt{(2} -\sqrt{(2} -\sqrt{(2} +\sqrt{(2} -\sqrt{(2}$
$+\sqrt{(2} +\sqrt{(2} +\sqrt{(2} +\sqrt{(2} -\sqrt{(2} -\sqrt{(2} -\sqrt{(2} +\sqrt{(2} -\sqrt{(2} -\sqrt{(2} +\sqrt{(2} -\sqrt{(2}$
$+\sqrt{(2} +\sqrt{(2} +\sqrt{(2} +\sqrt{(2} -\sqrt{(2} -\sqrt{(2} -\sqrt{(2} +\sqrt{(2} -\sqrt{(2} -\sqrt{(2} +\sqrt{(2} -\sqrt{(2}$
$+\sqrt{(2} +\sqrt{(2} +\sqrt{(2} +\sqrt{(2} -\sqrt{(2} -\sqrt{(2} -\sqrt{(2} +\sqrt{(2} -\sqrt{(2} -\sqrt{(2}\,]/2$

Sin 88° = ½ $\sqrt{(2} + \sqrt{(2}$ $\underline{+\sqrt{(2} +\sqrt{(2} +\sqrt{(2} +\sqrt{(2} -\sqrt{(2} -\sqrt{(2} -\sqrt{(2} +\sqrt{(2}}$
$\underline{-\sqrt{(2} -\sqrt{(2} +\sqrt{(2} -\sqrt{(2}}\,]$

[Please care the sign of <u>repetition</u> has <u>underlined</u>.]

Sin 3° (Crd. 6°)

This is odd integer angle. First three radicals will be non-repetitive. Then after there will be repeating pattern of nested radicals as:

Sin 3° = $\sqrt{(2} -\sqrt{(2} +\sqrt{(2}$
$\underline{+\sqrt{(2} +\sqrt{(2} -\sqrt{(2} -\sqrt{(2}}$
$+\sqrt{(2} +\sqrt{(2} -\sqrt{(2} -\sqrt{(2}$
$+\sqrt{(2} +\sqrt{(2} -\sqrt{(2} -\sqrt{(2}$
$+\sqrt{(2} +\sqrt{(2} -\sqrt{(2} -\sqrt{(2}$
$+\sqrt{(2} +\sqrt{(2} -\sqrt{(2} -\sqrt{(2}$
$+\sqrt{(2} +\sqrt{(2} -\sqrt{(2} -\sqrt{(2}$
$+\sqrt{(2} +\sqrt{(2} -\sqrt{(2} -\sqrt{(2}$
$+\sqrt{(2} +\sqrt{(2} -\sqrt{(2} -\sqrt{(2}$
$+\sqrt{(2} +\sqrt{(2} -\sqrt{(2} -\sqrt{(2}$
$+\sqrt{(2} +\sqrt{(2} -\sqrt{(2} -\sqrt{(2}$
$+\sqrt{(2} +\sqrt{(2} -\sqrt{(2} -\sqrt{(2}$
$+\sqrt{(2} +\sqrt{(2} -\sqrt{(2} -\sqrt{(2}$
$+\sqrt{(2} +\sqrt{(2}\,]/2$

Therefore,

Sin 3° = ½ $\sqrt{(2} -\sqrt{(2} +\sqrt{(2} + \sqrt{(2}$ $\underline{+\sqrt{(2} -\sqrt{(2} -\sqrt{(2}}\,]$

Sin 87° = ½ $\sqrt{(2} + \sqrt{(2} +\sqrt{(2}$ $\underline{+\sqrt{(2} +\sqrt{(2} -\sqrt{(2} -\sqrt{(2}}\,]$

[Please care the sign of <u>repetition</u> has <u>underlined</u>.]

PRECISE-REWRITTEN METHOD

Sin 4° (Crd. 8°)

This is even integer angle divisible by 4. First two radicals will be non-repetitive. Then after there will be repeating pattern of nested radicals as:

Sin 4° = √(2 -√(2

<u>+√(2 +√(2 +√(2 -√(2 -√(2 -√(2 +√(2 -√(2 -√(2 +√(2 -√(2 +√(2</u>

+√(2 +√(2 +√(2 -√(2 -√(2 -√(2 +√(2 -√(2 -√(2 +√(2 -√(2 +√(2

+√(2 +√(2 +√(2 -√(2 -√(2 -√(2 +√(2 -√(2 -√(2 +√(2 -√(2 +√(2

+√(2 +√(2 +√(2 -√(2 -√(2 -√(2 +√(2 -√(2 -√(2 +√(2 -√(2 +√(2

+√(2 +√(2 +√(2]/2

Sin 4° = √(2 -√(2 <u>+√(2 +√(2 +√(2 -√(2 -√(2 -√(2 +√(2 -√(2 -√(2 +√(2 -√(2 +√(2]</u>

Sin 86° = √(2 + √(2 + <u>√(2 +√(2 +√(2 -√(2 -√(2 -√(2 +√(2 -√(2 -√(2 +√(2 -√(2 +√(2]</u>

Alternative solution exists in case of Sin 86°.

[Please care the sign of <u>repetition</u> has <u>underlined</u>.]

Sin 5° (Crd. 10°)

This is odd integer angle. First three radicals will be non-repetitive. Then after there will be repeating pattern of nested radicals as:

Sin 5° = √(2 -√(2 +√(2

<u>+√(2 +√(2 -√(2</u>

+√(2 +√(2 -√(2

+√(2 +√(2 -√(2

+√(2 +√(2 -√(2 +√(2 +√(2 -√(2 +√(2 +√(2 -√(2 +√(2 +√(2 -√(2 +√(2 +√(2 -√(2 +√(2 +√(2 -√(2 +√(2 +√(2 -√(2 +√(2 +√(2 -√(2 +√(2 +√(2 -√(2 +√(2]/2

Sin 5° = ½ √(2 -√(2 +√(2 <u>+√(2 +√(2 -√(2]</u>

Sin 85° = ½ √(2 + √(2 +√(2 <u>+√(2 +√(2 -√(2]</u>

Sin 6° (Crd. 12°)

This is even integer angle (and not divisible by 4). Only the first radical will be non-repetitive. Then after there will be repeating pattern of nested radicals as:

Sin 6° = √(2
<u>-√(2 +√(2 +√(2 -√(2</u>
-√(2 +√(2 +√(2 -√(2
-√(2 +√(2 +√(2 -√(2
-√(2 +√(2 +√(2 -√(2
-√(2 +√(2 +√(2 -√(2 -√(2 +√(2 +√(2 -√(2 -√(2 +√(2 +√(2 -√(2
-√(2 +√(2 +√(2 -√(2 -√(2 +√(2 +√(2 -√(2 -√(2 +√(2 +√(2 -√(2
-√(2 +√(2 +√(2 -√(2 -√(2 +√(2 +√(2 -√(2 -√(2 +√(2 +√(2]/2

Sin 6° = ½ √(2 <u>-√(2 +√(2 +√(2 -√(2</u>]
Sin 84° = ½ √(2 - <u>-√(2 +√(2 +√(2 -√(2</u>]

Alternative solution exists in case of Sin 84°.

[Please care the sign of <u>repetition</u> has <u>underlined</u>.]

Sin 7° (Crd. 14°)

This is odd integer angle. First three radicals will be non-repetitive. Then after there will be repeating pattern of nested radicals as:

Sin 7° = √(2 -√(2 +√(2
<u>+√(2 -√(2 -√(2 +√(2 -√(2 +√(2 +√(2 +√(2 +√(2 -√(2 -√(2 -√(2</u>
+√(2 -√(2 -√(2 +√(2 -√(2 +√(2 +√(2 +√(2 +√(2 -√(2 -√(2 -√(2
+√(2 -√(2 -√(2 +√(2 -√(2 +√(2 +√(2 +√(2 +√(2 -√(2 -√(2 -√(2
+√(2 -√(2 -√(2 +√(2 -√(2 +√(2 +√(2 +√(2 +√(2 -√(2 -√(2 -√(2
+√(2 -√(2]/2

Sin 7° = ½ √(2 -√(2 +√(2 <u>+√(2 -√(2 -√(2 +√(2 -√(2 +√(2 +√(2 +√(2 +√(2 -√(2 -√(2 -√(2</u>]

Sin 83° = ½ √(2 + √(2 +√(2 <u>+√(2 -√(2 -√(2 +√(2 -√(2 +√(2 +√(2 +√(2 +√(2 -√(2 -√(2 -√(2</u>]

[Please care the sign of <u>repetition</u> has <u>underlined</u>.]

Sin 8° (Crd. 16°)

This is even integer angle divisible by 4. First two radicals will be non-repetitive. Then after there will be repeating pattern of nested radicals as:

Sin 8° = √(2 -√(2

±√(2 +√(2 -√(2 -√(2 -√(2 +√(2 -√(2 -√(2 +√(2 -√(2 +√(2 +√(2

+√(2 +√(2 -√(2 -√(2 -√(2 +√(2 -√(2 -√(2 +√(2 -√(2 +√(2 +√(2

+√(2 +√(2 -√(2 -√(2 -√(2 +√(2 -√(2 -√(2 +√(2 -√(2 +√(2 +√(2

+√(2 +√(2 -√(2 -√(2 -√(2 +√(2 -√(2 -√(2 +√(2 -√(2 +√(2 +√(2

+√(2 +√(2 -√(2]/2

Sin 8° = ½ √(2 -√(2 ±√(2 +√(2 -√(2 -√(2 -√(2 +√(2 -√(2 -√(2 +√(2 -√(2 +√(2 +√(2]

Sin 82° = ½ √(2 + √(2 ±√(2 +√(2 -√(2 -√(2 -√(2 +√(2 -√(2 -√(2 +√(2 -√(2 +√(2 +√(2]

Sin 9° – Sin 45°

Similar logic has used in the following integer angles in the sequence from 9° to 45° and their complementary angle (81° to 45°). More simple alternative than above rule of repetition has taken in few cases.

[Please care the sign of <u>repetition</u> has <u>underlined</u>.]

$A°$	$\sin A°$

9   ½ √(2 - √(2 + √(2 +√(2 - √(2]

10   ½ +√(2 - √(2 + √(2]

11   √(2 - √(2 + √(2 +√(2 - √(2 + √(2 + √(2 + √(2 + √(2 - √(2 - √(2 - √(2 + √(2 - √(2 - √(2]

12   ½ √(2 - √(2 + √(2 - √(2 - √(2 + √(2]

13   ½ √(2 - √(2 + √(2 -√(2 - √(2 + √(2 - √(2 - √(2 + √(2 - √(2 + √(2 + √(2 + √(2 + √(2 - √(2]

14   ½ √(2 -√(2 + √(2 - √(2 - √(2 + √(2 - √(2 + √(2 + √(2 + √(2 + √(2 - √(2 – √(2]

15   ½ √(2 - √(2 + √(2 - √(2]

16   ½ √(2 - √(2 + √(2 - √(2 - √(2 - √(2 + √(2 - √(2 - √(2 + √(2 - √(2 + √(2 + √(2 + √(2]

17   ½ √(2 - √(2 + √(2 -√(2+ √(2 - √(2 + √(2 + √(2 + √(2 + √(2 - √(2 - √(2 - √(2 + √(2 - √(2]

18   ½ +√(2 - √(2]

19   ½ √(2 - √(2 + √(2 -√(2 + √(2 - √(2 - √(2 + √(2 - √(2 + √(2 + √(2 + √(2 + √(2 - √(2 - √(2]

20   ½ √(2 -√(2 + √(2 -√(2 +√(2]

21   ½ √(2 -√(2 +√(2 -√(2 + √(2 +√(2 -√(2 -√(2 +√(2 +√(2 -√(2]

Easy alternative is ½ √(2 - √(2 + √(2 - √(2 +√(2 + √(2 - √(2 - √(2]

22   ½ √(2 -√(2 + √(2 - √(2 + √(2 + √(2 + √(2 + √(2 - √(2 - √(2 - √(2 + √(2 - √(2]

23	½ √(2 - √(2 - √(2 -√(2 + √(2 + √(2 + √(2 + √(2 - √(2 - √(2 - √(2 + √(2 - √(2 - √(2 + √(2]
24	½ √(2 - √(2 - √(2 - √(2 + √(2 + √(2]
25	½ √(2 - √(2 - √(2 -√(2 + √(2 + √(2]
26	½ √(2 -√(2 -√(2 -√(2 +√(2 -√(2 -√(2 +√(2 -√(2 +√(2 +√(2 +√(2 +√(2]
27	½ √(2 - √(2 - √(2 - √(2 + √(2]
28	½ √(2 - √(2 - √(2 - √(2 + √(2 - √(2 + √(2 + √(2 + √(2 + √(2 - √(2 - √(2 + √(2]
29	½ √(2 - √(2 - √(2 - √(2 - √(2 - √(2 + √(2 - √(2 - √(2 + √(2 - √(2 + √(2 + √(2 + √(2 + √(2]
30	½ √(2 - √(2]
31	½ √(2 - √(2 - √(2 - √(2 - √(2 + √(2 - √(2 + √(2 + √(2 + √(2 + √(2 - √(2 - √(2 - √(2 + √(2 - √(2 - √(2]
32	½ √(2 - √(2 -√(2 - √(2 - √(2 + √(2 - √(2 - √(2 + √(2 - √(2 + √(2 + √(2 + √(2 + √(2]
33	½ √(2 - √(2 - √(2 -√(2 - √(2 + √(2 + √(2 - √(2 - √(2 + √(2 + √(2]
34	½ √(2 -√(2 -√(2 +√(2 -√(2 +√(2 +√(2 +√(2 +√(2 -√(2 -√(2 -√(2 +√(2]
35	½ √(2 - √(2 - √(2 + √(2 - √(2 + √(2 + √(2 - √(2 + √(2]
36	½ √(2 - √(2 - √(2 +√(2 - √(2]
37	½ √(2 - √(2 - √(2 + √(2 - √(2 - √(2 - √(2 + √(2 - √(2 - √(2 + √(2 - √(2 + √(2 + √(2 + √(2]

38 $\frac{1}{2} \sqrt{(2 -\sqrt{(2 -\sqrt{(2 +\sqrt{(2 -\sqrt{(2 -\sqrt{(2 +\sqrt{(2 -\sqrt{(2 +\sqrt{(2 +\sqrt{(2 +\sqrt{(2 +\sqrt{(2 -\sqrt{(2}}}}}}}}}}}}}}$]

39 $\frac{1}{2} \sqrt{(2 - \sqrt{(2 - \sqrt{(2 + \sqrt{(2 - \sqrt{(2 - \sqrt{(2 + \sqrt{(2}}}}}}}}$]

40 $\frac{1}{2} \sqrt{(2 - \sqrt{(2 -\sqrt{(2 + \sqrt{(2 + \sqrt{(2}}}}}}$]

41 $\frac{1}{2} \sqrt{(2 - \sqrt{(2 - \sqrt{(2 + \sqrt{(2 + \sqrt{(2 - \sqrt{(2 - \sqrt{(2 - \sqrt{(2 + \sqrt{(2 - \sqrt{(2 - \sqrt{(2 + \sqrt{(2 - \sqrt{(2 + \sqrt{(2 + \sqrt{(2}}}}}}}}}}}}}}}}$]

42 $\frac{1}{2} \sqrt{(2 -\sqrt{(2 -\sqrt{(2 +\sqrt{(2 +\sqrt{(2}}}}}}$]

43 $\frac{1}{2} \sqrt{(2 - \sqrt{(2 - \sqrt{(2 + \sqrt{(2 + \sqrt{(2 + \sqrt{(2 - \sqrt{(2 - \sqrt{(2 - \sqrt{(2 + \sqrt{(2 - \sqrt{(2 - \sqrt{(2 + \sqrt{(2 - \sqrt{(2 + \sqrt{(2}}}}}}}}}}}}}}}}$]

44 $\frac{1}{2} \sqrt{(2 -\sqrt{(2 -\sqrt{(2 +\sqrt{(2 +\sqrt{(2 +\sqrt{(2 +\sqrt{(2 -\sqrt{(2 - \sqrt{(2 -\sqrt{(2 +\sqrt{(2 -\sqrt{(2 -\sqrt{(2 +\sqrt{(2}}}}}}}}}}}}}}$]

45 $\frac{1}{2} \sqrt{2}$

2.4 Exact Values of polygon chords

In the above sub-chapter we observed exact value of integer angles. In this sub-chapter, we shall discuss the exact length of few of the polygons using Precise-Rewritten method. Guru Euclid is father of polygonal geometry. No one Himafter developed such huge contribution on polygon yet. This sub-chapter is in His respect.

For any regular polygon, there is pattern for chord or length of a side. If quotient on dividing 360 by the given angle is an integer, that is regular polygon for this purpose. For example

360°/1° = 360 (this is 360-gon, 1° has exact value in pattern)

360°/18.94737°= 19 (this is 19-gon, 18.94737° has exact value in pattern)

=360°/0.125874125874126°= 2860 (this is 2860-gon, 0.125874125874126° has exact value in pattern)

As a result, angle which is 360°/n [where n is a natural] has exact value of trigonometry. Following is the length of a side of regular polygon up to n= 24.

N-gon	Each side exact length
3	$\sqrt{2 + \sqrt{2 - \sqrt{2}}}$
4	$\sqrt{2}$
5	$\sqrt{2 - \sqrt{2 - \sqrt{2 + \sqrt{2}}}}$
6	$\sqrt{2 - \sqrt{2}}$
7	$\sqrt{2 - +\sqrt{2 - \sqrt{2 - \sqrt{2}}}}$
8	$\sqrt{2 - \sqrt{2}}$

9. $\sqrt{(2 - \sqrt{(2 + \sqrt{(2 - \sqrt{(2 + \sqrt{(2} + \sqrt{(2} - \sqrt{(2}}}}}}]$

10. $\sqrt{(2 - \sqrt{(2} + \sqrt{(2}}]$

11. $\sqrt{(2 - \sqrt{(2 + \sqrt{(2} - \sqrt{(2} - \sqrt{(2} - \sqrt{(2} + \sqrt{(2}}}]$

12. $\sqrt{(2 - \sqrt{(2 + \sqrt{(2 - \sqrt{(2}}}}]$

13. $\sqrt{(2 - \sqrt{(2 + \sqrt{(2} - \sqrt{(2} - \sqrt{(2} + \sqrt{(2} - \sqrt{(2} + \sqrt{(2}}}]$

14. $\sqrt{(2 - \sqrt{(2 + \sqrt{(2} - \sqrt{(2} - \sqrt{(2} + \sqrt{(2} - \sqrt{(2} - \sqrt{(2}}}]$

15. $\sqrt{(2 - \sqrt{(2 + \sqrt{(2} - \sqrt{(2} - \sqrt{(2} + \sqrt{(2}}}]$

16. $\sqrt{(2 - \sqrt{(2 + \sqrt{(2}}}]$

17. $\sqrt{(2 - \sqrt{(2 + \sqrt{(2 + \sqrt{(2 - \sqrt{(2} + \sqrt{(2} + \sqrt{(2} + \sqrt{(2}}}}}}]$

18. $\sqrt{(2 - \sqrt{(2} + \sqrt{(2} + \sqrt{(2}}]$

19. $\sqrt{(2 - \sqrt{(2 + \sqrt{(2 + \sqrt{(2} - \sqrt{(2} + \sqrt{(2} - \sqrt{(2} - \sqrt{(2} - \sqrt{(2} + \sqrt{(2} + \sqrt{(2}}}}]$

20. $\sqrt{(2 - \sqrt{(2 + \sqrt{(2 + \sqrt{(2} - \sqrt{(2}}}}]$

21. $\sqrt{(2 - \sqrt{(2 + \sqrt{(2 + \sqrt{(2} - \sqrt{(2} + \sqrt{(2} - \sqrt{(2} + \sqrt{(2} + \sqrt{(2}}}}]$

22. $\sqrt{(2 - \sqrt{(2 + \sqrt{(2} + \sqrt{(2} - \sqrt{(2} - \sqrt{(2} - \sqrt{(2}}}]$

23. $\sqrt{(2 - \sqrt{(2 + \sqrt{(2} + \sqrt{(2} - \sqrt{(2} - \sqrt{(2} \sqrt{(2} + \sqrt{(2} - \sqrt{(2} + \sqrt{(2} - \sqrt{(2} - \sqrt{(2} + \sqrt{(2}}}]$

24. $\sqrt{(2 - \sqrt{(2 + \sqrt{(2 + \sqrt{(2 - \sqrt{(2}}}}]$

PRECISE-REWRITTEN METHOD

How to find the exact value of a polygon

Finding length of each side of a polygon is simple use of Precise-Rewritten method.

Firstly, find the angle for the n-gon using[6] $180/n$.

Secondly, find the chord length using Precise-Rewritten method.

Thirdly, carefully observe and make formula for repeated nested radicals.

For example for 200-gon. Its central angle will be $360/200 = 1.8°$. But our calculation formula is Sine-based, so we use $180/200 = 0.9°$

Using Precise-Rewritten through binary-technique

Sin $0.9°$ = √(2 -√(2 +√(2 +√(2

+√(2 +√(2 +√(2 -√(2 -√(2 -√(2 -√(2 +√(2 +√(2 -√(2

+√(2 +√(2 +√(2 -√(2 -√(2 -√(2 -√(2 +√(2 +√(2 -√(2

+√(2 +√(2 +√(2 -√(2 -√(2 -√(2 -√(2 +√(2 +√(2 -√(2

+√(2 +√(2 +√(2 -√(2 -√(2 -√(2 -√(2 +√(2 +√(2 -√(2

+√(2 +√(2 +√(2 -√(2 -√(2 -√(2 -√(2 +√(2 +√(2 -√(2

+√(2 +√(2]/2

So, Chord $1.8°$ = Side length of 200-gon = √(2 -√(2 +√(2 +√(2

+√(2 +√(2 +√(2 -√(2 -√(2 -√(2 -√(2 +√(2 +√(2 -√(2]

---

[6] Here, we use $180°$ instead of $360°$. The reason is Precise-Rewritten method delibarately design for Sin A which is half of Crd. 2A. For Crd. A, we need word down one more step of calculation.

2.5 Proof for above values are exact

We have seen few of known chords or values of Sine has changed into nested radical of 2. Does these nested radicals represent the real known values is one of the crucial question. Let us take few example of proof of accuracy of results from Precise-Rewritten method.

For proving easily, we take the known chords to fit above examples.

Chord 36°

Sin 18° is half of chord 36° (a). Let us proof above result of chord 36° (double of Sin 18°) is same as our known chord $(\sqrt{5}-1)/2$.

Under Precise-Rewritten method, $\sin 18° = \sqrt{2 - \sqrt{2 + \sqrt{2 - \sqrt{2 + \sqrt{2 - \sqrt{2 + \sqrt{2 - \sqrt{2 + \sqrt{2 - \sqrt{2 + \sqrt{2 - \sqrt{2 + \sqrt{2 - \sqrt{2 + \sqrt{2 - \sqrt{2 + \sqrt{2 - \sqrt{2 + \sqrt{2 - \sqrt{2 + \sqrt{2 - \sqrt{2 + \sqrt{2 - \sqrt{2 + \sqrt{2 - \sqrt{2 + \sqrt{2 - \sqrt{2 + \sqrt{2 - \sqrt{2 + \sqrt{2 - \sqrt{2 + \sqrt{2 - \sqrt{2 + \sqrt{2 - \sqrt{2 + \sqrt{2 - \sqrt{2}}}}}}}}}}}}}}}}}}}}}}}}}}}}}}}}}}}/2$

Therefore, chord 36° = $\sqrt{2 - \sqrt{2 + \sqrt{2 - \sqrt{2 + \sqrt{2 - \sqrt{2 + \sqrt{2 - \sqrt{2 + \sqrt{2 - \sqrt{2 + \sqrt{2 - \sqrt{2 + \sqrt{2 - \sqrt{2 + \sqrt{2 - \sqrt{2 + \sqrt{2 - \sqrt{2 + \sqrt{2 - \sqrt{2 + \sqrt{2 - \sqrt{2 + \sqrt{2 - \sqrt{2 + \sqrt{2 - \sqrt{2 + \sqrt{2 - \sqrt{2 + \sqrt{2 - \sqrt{2 + \sqrt{2 - \sqrt{2 + \sqrt{2 - \sqrt{2 + \sqrt{2 - \sqrt{2}}}}}}}}}}}}}}}}}}}}}}}}}}}}}}}}}}}$.

If we carefully observed, chord 36° is

$a = \sqrt{2 - \sqrt{2 + \cdots}}$. We knew chord 36° (a) = $(\sqrt{5}-1)/2$.

Hence, we need to prove, $\sqrt{2 - \sqrt{2 + \cdots}} = (\sqrt{5}-1)/2$.

Squaring both sides of $a = \sqrt{2 - \sqrt{2 + \cdots}}$, is $a^2 = 2 - \sqrt{2 + \sqrt{2 - \cdots}} = 2 - \sqrt{2 + a}$

Or, $a^2-2 = \sqrt{(2+a)}$

Or, $(a^2-2)^2 = (2+a)$

Or, $a^4 - 4a^2 + 4 - 2 - a = 0$

Or, $a^4 - 4a^2 - a + 2 = 0$

Or, $a^4 - 2a^3 + 2a^3 - 4a^2 - a + 2 = 0$

Or, $a^3(a-2) + 2a^2(a-2) - (a-2) = 0$

Or, $(a-2)(a^3 + 2a^2 - 1) = 0$

So, $a^3 + 2a^2 - 1 = 0$ [Since, $a = 2$ is more than 1 is not expected chord]

Taking, $a^3 + 2a^2 - 1 = 0$

Or, $a^3 + a^2 + a^2 + a - a - 1 = 0$

Or, $a^2(a+1) + a(a+1) - 1(a+1) = 0$

Or, $(a+1)(a^2 + a - 1) = 0$

So, $a^2 + a - 1 = 0$ [Since, $a = -1$ is not a chord]

Using binomial solution,

Here, $a \;\; = ½[-1 \pm \sqrt{((-1)^2 - 4 \times 1 \times (-1))}]$

$\qquad\qquad = ½[-1 \pm \sqrt{(1+4)}]$

$\qquad\qquad = ½[-1 \pm \sqrt{5}]$

$\qquad\qquad = ½[\sqrt{5}-1]$ [Since, ½$[-\sqrt{5}-1]$ is not a chord]

Hence, Chord $36°$ $(a) = \sqrt{(2 - \sqrt{(2 + \cdots}}) = (\sqrt{5}-1)/2$

Chord 60°

We knew Chord 60° is 1 and its half is Sin 30°.

For Sin 30°, we have taken $\sqrt{2 - \sqrt{2}}$ as chord of 60° under Precise-Rewritten method. We need to prove $\sqrt{2 - \sqrt{2}} = 1$ for accuracy of Precise-Rewritten method.

Here, a = $\sqrt{2 - \sqrt{2}}$

First of all, using sum of nested radicals for n = 2 with subtraction formula (adjustment of Guru Ramanujan) given in sub-chapter 1.4,

$$\begin{aligned}
\text{Sum S} &= \tfrac{1}{2}\left[-1+\sqrt{1+4n}\right] \\
&= \tfrac{1}{2}\left[-1+\sqrt{1+4\times 2}\right] \\
&= \tfrac{1}{2}\left[-1+\sqrt{1+8}\right] \\
&= \tfrac{1}{2}\left[-1+3\right] \\
&= 1
\end{aligned}$$

Sum S is our destination chord for 60°.

Therefore, Crd. 60° = 1.

======

For those, who have little knowledge about denesting above, Guru Ramanujan's denesting radical formula is as follows:

$$\sqrt{n - \sqrt{n - \sqrt{n - \ldots}}}$$

$$= \frac{-1+\sqrt{1+4n}}{2}.$$

1. For n=2, chord (a) = $\frac{-1+\sqrt{1+4\times 2}}{2} = \frac{-1+3}{2} =$

======

Alternatively, we can compute Crd. 60° using algebraic equation as;

$a = \sqrt{2 - \sqrt{2}}$

Squaring both sides,

$a^2 = 2 - \sqrt{2} = 2 - a$

Or, $a^2 + a - 2 = 0$

Or, $a^2 - a + 2a - 2 = 0$

Or, $a(a - 1) + 2(a - 1) = 0$

Or, $(a - 1)(a + 2) = 0$

Or, $a = 1$ or -2

A chord cannot be negative as well as more than 1. Therefore, we have single option of $a = 1$.

2.6 Constants for some nested radicals

Here are some of the nested radicals with values. Few of them are common for more than one angle too. These values are based on the formula and concept given in sub-chapter 1.4 Sum of Nested radicals.

The constants are given upto 14-digits of accuracy. One need to more accuracy need to further calculate.

a. Repeating nested radical $+\sqrt{(2 - \sqrt{(2 + \sqrt{(2]}}}$ has used in Crd. 20 ° and Crd. 10° and represents 1.53208888623796 …. . To determine this value, firstly, compute the value of $\sqrt{(2 - \sqrt{(2 + \sqrt{2}))}}$. This will be 'n'. Use the sum of nested radical formula, $S = ½ [1+\sqrt{(1+4n)}]$ for infinite occurrences.

b. Repeating nested radical $-\sqrt{(2 + \sqrt{(2 + \sqrt{(2]}}}$ $-\sqrt{(2 + \sqrt{(2 + \sqrt{(2}}}$ has used in Crd. 40°. Its decimal value for 14-digits accuracy is -1.90743390149645 ….. therefore, Sin 20 will be ½ $\sqrt{(2 - \sqrt{(2 + \sqrt{0.0925660985035461]}}}$

c. For 3° and 6°, repeating nested radical $+\sqrt{(2 +\sqrt{(2 -\sqrt{(2 -\sqrt{(2]}}}}$ has used with constant value of 1.8270909152852 ….

d. For 12° and 24°, nested radical $+\sqrt{(2 - \sqrt{(2 - \sqrt{(2 + \sqrt{(2}}}}$ has used with constant value of 1.33826121271772 ….

e. We can prepare the constant as required to use for conversion of exact value of chords from above exact values of integer angles or exact values of polygon chapter.

PRECISE-REWRITTEN METHOD

Request for the Readers

During the study for exact trigonometric values, I found few new items of values. I am working for their status and their interlinks with other parameter. I request professional readers for additional work on it. Their professional knowledge may contribute our mathematics universe more than my amateur hobby.

Case I: **Polygon angles are 3-based**

Length of side of polygon (and all angles) are routed through a radical of $\sqrt{(3-x)}$. this radical 3 is nothing but just there are 3 cuts up to 180° by a compass in the circumference. This leads the polygon are number of cut in the circumference based. There are 3(*) numbers are constants as required to fulfill the requirement of $\sqrt{(3-x)}$, yet to be proved.

N-gon	x $\sqrt{(3-x)}$	Length of Side	Decimal
1	3	$\sqrt{(3-0)}$	0
2	-1	$\sqrt{(3+1)}$	2
3	0	$\sqrt{(3-0)}$	1.732051
4	1	$\sqrt{(3-1)}$	1.414214
5	φ	$\sqrt{(3-\varphi)}$	1.175571
6	2	$\sqrt{(3-2)}$	1
7	2.2469796*	$\sqrt{(3-2.2469796)}$	0.867767
8	$\sqrt{2}+1$	$\sqrt{(3-\sqrt{2}-1)}$	

		= √(2-√2)	0.765367
9	2.532089*	√(3-2.532089)	0.68404
10	1+ φ	√(3-1+ φ)	
		=√(2+ φ)	0.618034
11	2.682507*	√(3-0)	0.563465
12	√3+1	√(3-√3-1)	
		=√(2-√3)	0.517638
13	2.792893		
		= ½ (7-√2)	√(3-½
(7-√2))	0.45509		

Case II: Tridecagon (13-gon) is 7-based

In the table above, tridecagon (13-gon) has a constant of 2.792893 (more accurate 2.79289321881345) which is ½ (7-√2). Well-known Euclidian pentagon and decagon are 5-based; triangle or hexagon are 3-based and tetragon, octagon, or hexadecagon are 2-based. Above table shows, there is something, which is 7-based or may be 6 or 8-based too.

The reason for finding any new base is opened by trigonometry itself. Chord or Sine of an angle may be expressed in various ways. For example chord $60°$, normally is, 1 and √3. Therefore, Value of Sin $60°$ is √3/2. In the other part, it may be expressed, if the context requires, as:

Sin $60°$= $\sqrt{4-1}$/2 (supplementary chord formula)

Sin $60°$= $\sqrt{5-\sqrt{4}}$/2 (Phi-route formula)

Sin $60°$= $\sqrt{6-\sqrt{9}}$/2 (9-gon-route formula)

We used another method of presentation in Precise-Rewritten method.

Therefore, there might be formula having 7-base or 8-base or 9-base for exact values. Need not to say, those will be additional formula over Precise-Rewritten method of trigonometry.

Case III   **2.53208888623796 …. is a usable constant**

In above table, there are three (*). The values were back computed and still unknown for other link. Based on Precise-Rewritten method, above values can be calculated. One of the value 2.53208888623796 .. .. .. is equivalent to Sin 20°/Sin 60°. There may be interlink relation or Sin 20°/Sin 60° is just an arbitrary matching is still unknown. Once we conclude the reason of relation of Sin 20°/Sin 60° is 2.53208888623796 .. .. .. (say as β), we can compute all the trigonometric values using classical methods for integer angles.

Case IV   **Chord-Triangle or Sine- Triangle**

There are some triangles (or quadrilaterals etc.) which have chord as each side or length of Sine as length of sides. For example, in 60° - 60° -60° triangle, each side may be √3. This is the case of Chord-triangle. In such triangle, each side may be √3/2. This is the case of Sine-triangle.

In the geometry, there are numerous combination of chord-triangle or Sine-triangle. The relation amongst Chord-triangles or Sine-Triangles are unknown yet.

***

www.ingramcontent.com/pod-product-compliance
Lightning Source LLC
Chambersburg PA
CBHW070332190526
45169CB00005B/1862